SpringerBriefs in Microbiology

Extremophilic Bacteria

Series editors

Sonia M. Tiquia-Arashiro, Dearborn, MI, USA
Melanie Mormile, Rolla, MO, USA

More information about this series at http://www.springer.com/series/11917

Lesley-Ann Giddings · David J. Newman

Bioactive Compounds from Terrestrial Extremophiles

 Springer

Lesley-Ann Giddings
Natural Products Branch,
    Developmental Therapeutics Program
NCI
Frederick, MD
USA

David J. Newman
Natural Products Branch,
    Developmental Therapeutics Program
NCI
Frederick, MD
USA

ISSN 2191-5385   ISSN 2191-5393   (electronic)
SpringerBriefs in Microbiology
ISBN 978-3-319-13259-4   ISBN 978-3-319-13260-0   (eBook)
DOI 10.1007/978-3-319-13260-0

Library of Congress Control Number: 2014955318

Springer Cham Heidelberg New York Dordrecht London

Printed on acid-free paper

Springer International Publishing AG Switzerland is part of Springer Science+Business Media
(www.springer.com)

Contents

Bioactive Compounds from Terrestrial Extremophiles 1
1   Introduction 2
2   Thermophiles 5
3   Psychrophiles 15
4   Acidophiles. 22
     4.1   Berkeleydione, Berkeleytrione, and Berkeleyones A–C 24
     4.2   Berkelic Acid 24
     4.3   Berkeleyacetals A–C 26
     4.4   Berkeleyamides A–D 26
     4.5   Berkazaphilones A–C 27
     4.6   Other Metabolites 27
5   Alkaliphiles. 32
6   Halophiles. 41
7   Terrestrial Extremophiles Living in Mutualistic Environments? 51
8   Other Extremophiles. 62
     8.1   Endoliths. 63
     8.2   Other Rock-associated Microorganisms 63
     8.3   Xerophiles. 64
     8.4   Toxitolerant Extremophiles. 68
     8.5   Metallotolerant Microbes 69
     8.6   Unclassifiable Extremophiles 70
9   Summary and Concluding Remarks. 74

References 77

Bioactive Compounds from Terrestrial Extremophiles

Abstract Since the 1960s, bioactive secondary metabolites have been isolated and structurally characterized from eubacteria, archaea, and fungi, microorganisms that have adopted strategies to grow in extreme terrestrial environments. This book highlights a myriad of natural products that have been isolated from terrestrial extremophiles, which the authors define as microorganisms found on land as well as in rivers and salterns but exclude those that dwell in water beyond seashores. The isolation and bioactivity of these molecules from the following extremophiles are described: thermophiles, psychrophiles, acidophiles, alkaliphiles, halophiles, xerophiles, epi- or endophytes, as well as metallotolerant, radioresistant, and unclassifiable microorganisms. Quite a few of these secondary metabolites have entered (pre)clinical trials, and their current status in the drug evaluation process is discussed. The authors also describe the isolation of natural products solely produced by endophytes via mutualistic interactions between the host and the crop and raise the argument of whether these metabolites are produced as a result of mutualistic interactions required for survival or epigenetic manipulations of microorganisms cohabiting within the same niche. This book concludes with a discussion of some of the challenges to accessing environments with significant biodiversity and the notion that as ecological niches and biotechnology evolve, extremophiles will also evolve to be continuous, untapped sources of novel drugs and drug leads.

Keywords Extremophiles · Terrestrial · Anticancer   agents · Anti-infectives · Prokaryote · Eukaryote · Antioxidants · Clinical drugs · Microorganisms · Natural products · Fungi · Eubacteria · Archaea · Thermophiles · Psychrophiles · Acidophiles · Alkaliphiles · Halophiles · Toxitolerant · Epiliths · Endoliths · Xerophiles · Epi- or endophytes · Metallotolerant · Radioresistant microorganisms

---

Data on the biological activity of most compounds are provided. If the values were reported in terms of weight per volume, the units were left unchanged. However, to compare biological data of molecules of different molecular weights, we converted units in terms of molarity for true comparisons to be made.

© The Author(s) 2015   1
L.-A. Giddings and D.J. Newman, *Bioactive Compounds from Terrestrial
Extremophiles*, Extremophilic Bacteria, DOI 10.1007/978-3-319-13260-0_1

1 Introduction

Over the last several decades, scientists have come to realize that life exists wherever there is a trace of water as well as in dry areas inhabited by fungi in the form of spores (Ball and Stillinger 1999). Microorganisms, such as archaea, eubacteria, and eukarya (fungi), have been isolated from environments with major chemical and physical barriers to support life. These organisms are classified as extremophiles, which are generally defined as single-celled organisms from either *Archaea* or *Prokarya* as well as *Eukarya* (i.e., fungi) that have evolved to live and reproduce optimally in extreme ranges of key environmental variables.

The definition of an "extreme" environment is anthropogenic, as from the point of view of the organism, the environment is considered to be "normal" for growth and reproduction (Horikoshi and Bull 2011; Zhang et al. 2013a). Scientific evidence suggests that Earth was initially a hot, dry, and anoxic hydrothermal vent. Thus, one could speculate that the earliest environment on Earth was "normal." For example, almost 2 billion years ago, the volatile gases trapped within the Earth's core, heated by radioactive decay, were released over time via volcanic outgassing to form a second atmosphere (Kasting 1993). The atmosphere during the period of 4.4–3.8 Gya is speculated to have been dominated by steam, until the Earth's temperature dropped below 100 °C and water condensed to form oceans. From this perspective, humans and other microorganisms inhabit extreme environments. An alternative perspective is that all physical factors are on a continuum, and those that make it more difficult for an organism to survive create extreme environments (Rothschild and Mancinelli 2001). Extreme environments are considered to be hostile because they are "nonmesophilic" (either hot or cold; e.g., geysers), oligotrophic, or subject to osmotic stress, high salt concentrations, extreme pH, severe radiation, metals, and high pressures. To survive in these environments, cells have to live at certain temperatures as well as have mechanisms to regulate intracellular pH values, compositions of solutes, biochemical redox reactions, and the production of other biomolecules, and repair DNA damage. Some factors, such as radiation and oxygen, destroy biomolecules by damaging DNA, proteins, and lipids. For the purposes of this book, we have chosen to use the latter perspective to define extreme environments.

There are extreme-tolerant microorganisms that survive in extreme environments but reproduce optimally under moderate conditions. Other microorganisms can survive under extreme conditions in a dormant state but cannot grow and reproduce in these environments; however, they will not be considered in this book. Several representative examples of secondary metabolites produced from the following types of extremophiles, many of which fall into multiple categories, will be discussed: thermophiles, psychrophiles, acidophiles, alkaliphiles, halophiles, endoliths, hypoliths, xerophiles, epi- or endophytes (symbionts), as well as metallotolerant and radioresistant microorganisms. Because extremophiles typically fall within multiple categories, these microorganisms will be discussed under the most dominant environmental factor. Microorganisms that also grow and reproduce in unusual ecological niches, such as terrestrial insects, plants, fecal matter, the animal gut, or marine invertebrates, will also be discussed.

The distribution of extreme environments throughout the world

Extreme environments continuously arise due to the formation of endorheic basins, plate tectonic activity, the dynamic cryosphere, evolutionary mutualistic associations, and man-made changes to the environment (Horikoshi and Bull 2011). Endorheic basins are essentially drainage basins with no outflow of water (typically high in saline) with a salt lake or pan at the bottom of the basin. The formation of volcanoes, mid-ocean ridges, mountains, and deep-sea trenches as well as other geothermal phenomena can arise from plate tectonic activity, which is the colliding of the boundaries of two tectonic plates. The environments that result from these changes are typically high in pH, pressures, heat, metal content, and dissolved gases.

The cryosphere is composed of ice caps, glaciers, ice sheets, and permafrost and considered to be dynamic due to the constant influence of precipitation, ocean circulation, and hydrology. New ecosystems are currently evolving over geological timescales (e.g., subterranean rock formation) as well as those that result from human activity, such as oil spills, deep-well drilling for water, industrial pollution, or radioactivity contamination due to a power plant failure. Additionally, mutualistic associations between a microbe and host, such as marine invertebrates and plants, create unique ecological niches in which unique secondary metabolites are produced. Altogether, these dynamic or accidental changes create new unexplored ecosystems.

Many extreme ecosystems remain unexplored due to the costs and tremendous effort required to obtain samples from these extreme environments. However, since the mid-1970s, the number of extremophiles reported has increased to the extent that there are dedicated journals, scientific societies, and international symposia dedicated to the study of these organisms. Less than 5 % of all microbes, including those with examined biosynthetic machinery, can be cultivated under standard laboratory conditions, leaving the remaining 95 % to be isolated and cultured. Thus, the extreme biosphere has now become of significant interest as it represents and untapped source for biocatalysts and novel new natural products.

Early history of extremophiles

The first reported extremophile was a thermophile isolated in the 1860s, and it took over 100 years for the term "extremophiles" to be accepted to describe organisms that thrive in extreme environments (Pikuta et al. 2007). R.D. Macelroy first coined this term in his 1974 paper entitled "Some comments on the evolutions of extremophiles" (Macelroy 1974). Researchers first isolated and acknowledged the existence of extremophiles in the mid-to-late twentieth century. In 1965, the microbiologist Thomas Brock discovered a new form of eubacteria, *Thermus aquaticus*, from the thermal vents of Yellowstone National Park. Three years later, Koki Horikoshi isolated eubacteria growing in alkaline environments and has continued to identify more "alkaliphiles" and research their physiology.

Extremophiles have attracted significant attention due to their molecular evolution, which has improved their survival and molecular stability. Moreover, these microorganisms have been commercially exploited for their production of enzymes or "extremozymes," which have applications in several industrial processes, such as protein-degrading additives in detergents. Roughly 30 % of all the enzymes commercially produced in the world are used in the process of making detergents (Gerday and Glansdorff 2007). However, with the development of new molecular and genomic tools, the use of other small molecules produced by extremophiles is now being explored.

Extremophiles as new drug sources

Of all Food and Drug Administration (FDA)-approved drugs, roughly 50 % are derived from natural products or derivatives, which are most likely produced by microorganisms or symbionts (Giddings and Newman 2013). The secondary metabolites biosynthesized by microbes have been a major source of new drugs dating back to World War II (1939–1945), when the demand for penicillin and other new antibiotics led to major pharmaceutical companies forming large research programs centered on the discovery of natural products. This time period marked the beginning of the "Great Antibiotic Era," during which over a hundred pharmaceuticals from microbes were commercialized. Most of these drugs were produced by actinomycetes inhabiting terrestrial environments. From 1950 to 1990, pharmaceutical companies invested approximately 10 billion dollars each year for microbe-based drug discovery. During this period, academic laboratories and government agencies, including the National Cancer Institute (NCI), also assisted pharmaceutical companies and spearheaded research programs that led to the discovery of many natural products, which are either currently used in the clinic or in active development.

However, since the 1990s, large pharmaceutical companies have deemphasized natural product discovery due to high rediscovery rates from terrestrial sources and the advent of high-throughput screening programs based on molecular targets and combinatorial chemistry. These research programs were projected to speed up the drug development process and reduce costs. However, instead of creating large libraries, combinatorial chemistry is now used to build small, focused collections that mimic the core scaffolds of natural products, specifically their three-dimensional architectures, high density of functional groups, and the number of stereogenic centers and regiochemical constraints. Even with this shift in focus, there is still a shortage of lead compounds entering clinical trials. Approximately 50 % of all small molecules that have been approved by the FDA from 2000 to 2006 were natural products instead of new chemical entities derived from combinatorial chemistry.

To fully exhaust all possible chemical scaffolds from microorganisms, new niches need to be explored in addition to the development of new biochemical assays and cultivation methods. A diverse array of extremophiles, including eubacteria, cyanobacteria, some algae, and yeast, have been isolated from extreme conditions and are now being exploited for their secondary metabolites with new molecular frameworks. These organisms have adopted strategies to protect themselves against extreme environments. This book describes the unique, diverse bioactive secondary metabolites produced from microorganisms in extreme terrestrial environments, which we are defining as any microorganism that lives above the shores of the world's oceans. Extremophiles represent one of the last unexplored frontiers of drug discovery, as these unique environments often lead to the production of novel small molecules belonging to a wide array of unusual structural classes with novel biological activities. Importantly, these metabolites are stably produced by extremophiles within their distinctive niches and may serve as lead compounds in the development of novel drugs that are valuable to both animals and humans.

2 Thermophiles

The term "thermophile" originates from the Greek words "philos" (loving) and "thêrme" (heat). Thermophiles are microorganisms that inhabit hot environments at temperatures ranging from approximately 61–79 °C. Hyperthermophiles grow optimally at temperatures above 80 °C, and moderately thermophilic microbes grow between the temperatures of 50 and 60 °C. Thermophiles live in a wide range of high-temperature environments, including hot springs, bioreactors, deep oil wells, geothermal plants, coal piles, compost heaps, deep-sea hydrothermal vents (which will not be discussed in this book), and other subsurface environments (e.g., gold mines). In order to survive in these high-temperature environments, thermophiles modify the lipid compositions (e.g., longer lengths of lipid acyl chains) of their membranes such that they are in a liquid crystalline state, allowing the cell to function and reduce proton permeation rates.

Terrestrial thermophiles are found all over the world and have been extensively studied in Kamchatka, New Zealand, Thailand, Tibet, Italy, Chilé, and the Yellowstone National Park in the United States, notably for their production of heat-stable enzymes. For example, the first source of DNA Taq polymerase was isolated from the *T. aquaticus* strain T01 from a geothermal area in Yellowstone National Park. This discovery has since aided in the development of useful tools in the fields of genetics, molecular biology, enzymology, and other several areas of thermophilic microbiology. Furthermore, the utilization of thermophiles has been undoubtedly advantageous in several industrial processes, including anaerobic fermentative processes for water treatment, fuel production, as well as sulfur removal from crude oil.

Eubacteria and *Archaea* are widespread throughout most thermal environments, whereas only a few species of fungi and algae thrive under such heated conditions. Within the past decade, metagenomic analyses have demonstrated the diverse communities of thermophiles that thrive in these hot environments, although only a few, novel secondary metabolites have been isolated from these organisms. Secondary metabolites, such as ether-linked lipids, nucleosides, and

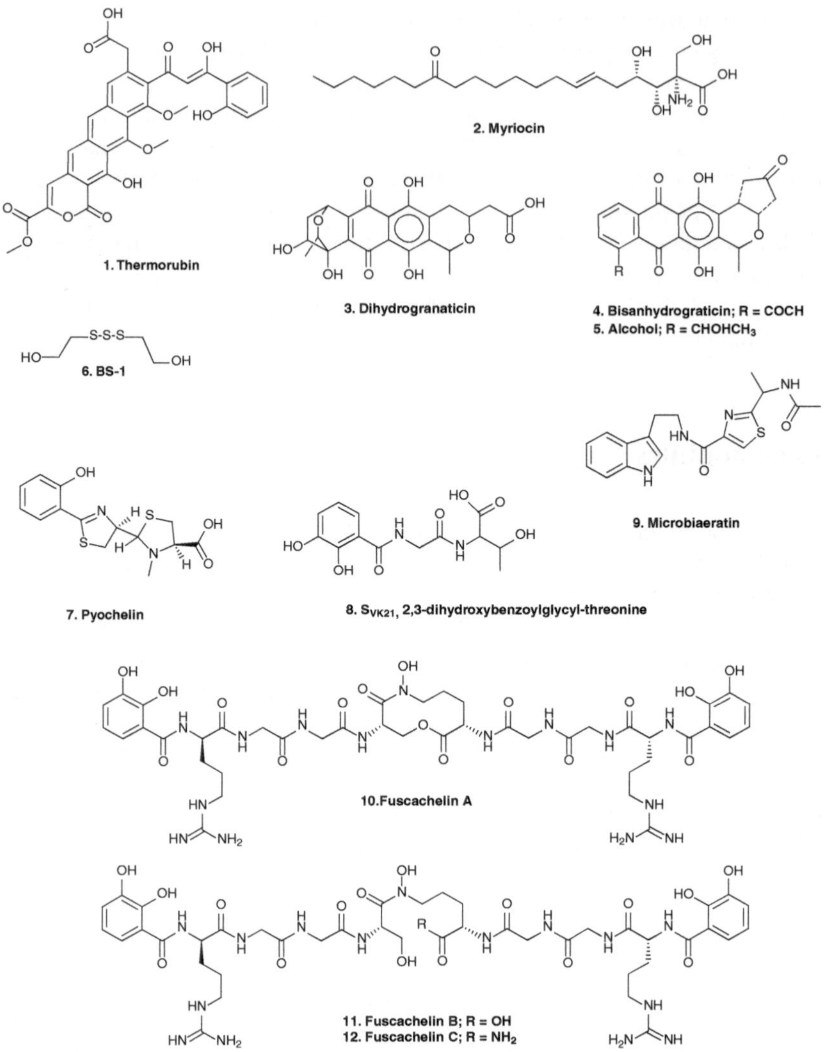

Fig. 1 Structures **1–12**

polyamines, are thought to play an important role in the survival of the microorganism at extreme temperatures.

The orange red pigment thermorubin was first isolated from a mildly thermophilic soil actinomycete, *Thermoactinomyces antibioticus*, in Pavia, Italy, in 1964 (Craveri et al. 1964). Its structure was initially elucidated in 1972 but later revised to be a 1*H*-2-anthro[2,3-*c*]pyran derivative (Johnson et al. 1980). Thermorubin **1** (Fig. 1) exhibited potent antimicrobial activity against Gram-positive eubacteria (*Staphylococcus aureus* ATCC 6538, MIC, 0.006 µg/ml; *S. aureus* TOUR, MIC, 0.05 µg/ml; *S. pyogenes* C 203 SKF 13400, MIC, 0.025 µg/ml; and *Streptococcus* *pneumonia* UC 41, MIC, 0.05 µg/ml), but lower activity against Gram-negative eubacteria (MIC values ranging from 0.1 to 25 µg/ml) and little to no activity against fungi and yeast (Cavalleri et al. 1985). When mice infected intraperitoneally with a species of *Streptococcus* were administered thermorubin intraperitoneally for 3 days at 3 mg/kg body weight, 100 % protection was observed with an LD_{50} of 300 mg/kg without cross-resistance with other antibiotics. Thus, compound **1** appeared to be a likely candidate for therapeutic development. However, when mice were administered the compound (up to 200 mg/kg) orally or subcutaneously, thermorubin was inactive, and some activity was observed when administered intravenously (ED_{50}, 50 mg/kg). Its poor solubility in physiological fluids and inactivation by serum proteins are speculated to account for its inactivity.

Thermorubin has been reported to be bacteriostatic and inhibit protein synthesis by inhibiting poly(U)-directed poly(Phe) synthesis and the binding of the initiator of protein synthesis, fmet-tRNA^{met}, to both subunits of the 70S ribosome. Although this molecule is structurally similar to tetracycline, thermorubin **1** has been reported to bind to a distinct location on the ribosome—at the interface between the small and large subunits formed by inter-subunit bridge B2a, which plays a role in the initiation of translation (Bulkley et al. 2012). Thus, thermorubin **1** most likely inhibits the initiation stage of protein synthesis. Thermorubin **1** has also been reported to be a potent aldose reductase inhibitor, reducing tissue sorbitol content in diabetic animals (Hayashi et al. 1995). Although thermorubin **1** is not very water soluble, this secondary metabolite may be useful in designing derivatives with the same mode of action and pharmacodynamics properties.

Several new bioactive metabolites have been isolated from thermophilic fungi. In 1972, an interesting sphingosine-like thermozymocidin was reported by Italian researchers following its isolation from the *Mycelia sterilia* (strain IPV F-4333) thermophilic fungus (Craveri et al. 1972; Aragozzini et al. 1972). This compound exhibited activity against a range of filamentous fungi and yeast. One year later, this same metabolite was isolated from the thermophilic fungus *Myriococcum albomyces* and renamed to be myriocin **2** (Fig. 1) (Bagli et al. 1973). Interestingly, myriocin **2** exhibited immunosuppressive activity against the appearance of plaque-forming cells in response to sheep red blood cells as well as T cell-dependent antibody production in mice by roughly 10-fold compared to that of cyclosporin A (Fujita et al. 1994). In addition, compared to cyclosporin A (minimum effective concentration, 10 mg/kg), a clinically prescribed immunosuppressant, myriocin **2** inhibited the generation of allo-reactive cytotoxic T lymphocytes in BALB/c mice after intraperitoneal or oral administration by 100-fold. Because myriocin **2** did not exhibit cytotoxic activity against several human leukemia cell lines in vitro, its antiproliferative activity in vivo appears to be specific to antigen-stimulated lymphocytes and not to growth inhibition, suggesting that myriocin **2** may have a different mode of action compared to that of the known immunosuppressants FK506 and cyclosporin A. In addition, myriocin **2** was later reported to inhibit a serine palmitoyltransferase (SPT) activity, which catalyzes the formation of 3-ketodihydrosphingosine, in murine cytotoxic T lymphocyte cell line (CTLL-2)-derived microsomes with an inhibition constant of 280 pM (Chen et al. 1999).

By evaluating myriocin **2** derivatives in CTLL-2 proliferation and SPT assays, the mammalian homologs of two yeast proteins involved in sphingolipid biosynthesis, LCB1 and LCB2s, were determined to be the proteins targeted by myriocin **2**. Based on these studies, sphingolipid derivatives can be used to probe the mechanisms required for sphingolipid homeostasis to prevent cell death.

In 1977, novel quinone antibiotics, such as dihydrogranaticin **3** and two other related anthraquinones (**4–5**) (Fig. 1), were reported to be produced from the thermophile *Streptomyces thermoviolaceus* subsp. *pigens* var WR-141 (Pyrek et al. 1977) and later from other nonthermophilic species of *Streptomyces* (e.g., *S. violaceoruber* and *S. lateritius*) (Snipes et al. 1979; Fleck et al. 1980). These metabolites were determined to be biosynthetic intermediates (some off-pathway) in the synthesis of the antibiotic granaticin A. Dihydrogranaticin **3** contains a unique 2-oxabicylo[2.2.2]oct-5-ene ring system fused to a napthoquinone ring with an open pyrano-γ-lactone ring. It displayed antimicrobial activity against *Bacillus cereus* (MIC, 0.3 μg/ml) and was later determined to inhibit viruses, mycobacteria, and both Gram-positive and Gram-negative eubacteria (Brimble et al. 1999). The other anthraquinones were determined to be bisanhydrogranaticin **4** and its alcohol derivative **5**, both of which are α-substituted quinizarins with a fused dihydropyranolactone bicyclic ring identical to that of granaticin. Bisanhydrogranaticin **4** and its alcohol derivative **5** exhibited antimicrobial activity against *B. cereus* with MIC values of 1.3 and 5 μg/ml, respectively. The identification and characterization of these compounds demonstrate the biosynthetic potential of thermophiles to produce several bioactive quinone derivatives via the expression of a single biosynthetic pathway.

After the 1970s, researchers began to screen more nonmesophilic microbes to find new bioactive chemical entities once they realized that terrestrial organisms had been screened several times, decreasing the possibility of finding new bioactive metabolites. As a result, an increasing number of reports were published on the identification of structurally unusual secondary metabolites from extremophiles. In 1991, a structurally interesting metabolite, bis(2-hydroxyethyl)trisulfide (BS-1) **6** (Fig. 1), originally reported in 1946 as a synthetic compound (Peppel and Signaigo 1946) used in the manufacturing of liquid polymers and vulcanization technology, was isolated from a natural source. BS-1 **6** was isolated from the thermophilic eubacterium *Bacillus stearothermophilus* UK563 (Kohama et al. 1991), structurally characterized (Kohama et al. 1992), and reported to activate macrophage-mediated cytotoxicity in P815 mouse mastocytoma cells (Kohama et al. 1993). This compound was later determined to induce the expression of mouse mitochondrial cytochrome b in mouse macrophage-like J774A.1 cells, which can lead to the generation or expression of nitric oxide, tumor necrosis factor-α (TNF-α), interleukin-1 (IL-1), and prostaglandin E2 (PGE2) to produce cytotoxic effects in cells (Kohama et al. 1996).

An increasing number of screening programs have been launched to find molecules with selective bioactivity. For example, Fish and coworkers screened a number of thermophilic and thermotolerant cyanobacteria and identified a thermotolerant species of *Phormidium* sp. isolated from a warm spring in the Azores that produced antimicrobial compounds active against a broad range Gram-positive (e.g., *S. aureus* NCIMB 3251) and Gram-negative (e.g., *Pseudomonas denitrificans* NCIMB 9496) heterotrophic eubacteria as well as yeast/fungi (*Candida albicans* 3153 A and *Cladosporium resinae*) (Fish and Codd 1994). Before this publication, no other cyanobacterium had been isolated and reported to have such a broad range of antimicrobial activity. Since this report, other research groups started screening thermophiles collected from various locations for antibiotics (Esikova et al. 2002; Alfredson et al. 2003).

The first whole-cell and mechanism-based screening program led by researchers at Waters Corporation, Montana Biotech, Alpha-Beta technology, Inc., and MycoLogics, Inc. was dedicated to solely isolating new antifungal agents from extremophiles from a number of thermal waters in Yellowstone National Park (Phoebe et al. 2001). Extracts of the biomasses and culture broths of 217 extremophiles from this Montana Biotech collection were screened against *C. albicans* and *Aspergillus fumigatus* to identify microbes with potent antifungal activity. This program led to the identification of the antifungal siderophore pyochelin **7** (Fig. 1), which was isolated from the thermophile *P. akbaalia*. This activity is not surprising as siderophores are commonly used as antimicrobial and iron-chelating agents in clinical medicine. Compound **7** was originally isolated from *P. aeruginosa*, the common cause of infection in animals (Cox et al. 1981), but the data corresponding to the antifungal activity of this pure compound were not reported.

Esikova and coworkers reported that an *n*-butanol extract of the thermophilic *B. licheniformis* VK21 exhibited antimicrobial activity against *B. megaterium* VKM41, *P. putida* I-97, *Staphylococcus sp.* SS1, and *Micrococcus luteus* E509 (Esikova et al. 2002). This thermophile was isolated from the thermal springs of the Kamchatka Peninsula, Russia. The authors later isolated a catecholic siderophore 2,3-dihydroxybenzoylglycyl-threonine (S_{VK21}, **8**) (Fig. 1) from *B. licheniformis* VK21 under growth iron-deficient conditions (Temirov et al. 2003). Prior to this report, there had only been a few reports on iron transport systems in thermophiles that utilize chelating metabolites (Guerry et al. 1997). Interestingly, S_{VK21} **8** is a fragment of the cyclic trilactone bacillibactin previously isolated from a genetic variant of the mesophilic *B. subtilis* (May et al. 2001). May and coworkers also isolated S_{VK21} **8** upon mutating the condensation domain required for cyclization to produce bacillibactin. Thus, *B. licheniformis* VK21 may have a similar mutation in an identical condensation domain. When the growth of *B. licheniformis* VK21 was inhibited by ETDA, the addition of compound **8** increased eubacterial growth, indicating that this compound is involved in iron transport in this thermophile. Although the crude extract of this thermophile was screened for antimicrobial activity, the bioactivity of S_{VK21} **8** against other microorganisms has not been reported. In 2007, the mildly thermophilic (found at 50 °C) *Microbispora aerate* strain IMBAS-11A was isolated from penguin excrements on Livingston Island, Antarctica, and reported to produce microbiaeratin **9** (Fig. 1) and the known alkaloid bacillibactin (Ivanova et al. 2007). Microbiaeratin **9** is an acetylated analog of TM-64, another thiazole-containing alkaloid isolated from the mildly thermophilic eubacterium *Thermoactinomyces* strain TM-64 (Ōmura et al. 1975;

Onda and Konda 1978). Compound **9** exhibited very weak antiproliferative and cytotoxic effects against L-929 mouse fibroblast cells as well as human leukemia K-562 cells and HeLa cervical cancer cells, but no antimicrobial activity against various fungi, microalgae, and Gram-positive and Gram-negative eubacteria.

One year later, new chemical entities were identified from thermophiles when the genome of the model thermophile *Thermobifida fusca* was sequenced. Dimise and coworkers used genome mining to identify an orphan gene cluster involved in the biosynthesis of the novel nonribosomal peptide siderophores, fuscachelins (Dimise et al. 2008). These 10-membered cyclic depsipeptides were the first secondary metabolites to be identified in *T. fusca*. Fuscachelins A–C (**10–12**) (Fig. 1) structurally differ by fuscachelin A **10** having a macrolactone ring, whereas fuscachelins B and C (**11–12**) lack this ring but possess carboxylic acid and amide moieties, respectively. These metabolites chelate ferric ions via the two terminal catechol and internal hydroxamate groups. Notably, these were first secondary metabolites to be identified from a thermophile via genome mining, demonstrating the vast potential of finding more diverse bioactive metabolites with the increasing availability of genome sequences. However, no other bioactivity has been reported for these siderophores.

Different screening tools, such as intact-cell desorption–ionization on silicon mass (DIOS) mass spectrometry (ICD-MS) and LC–solid-phase extraction–NMR (LC-SPE-NMR), have also been utilized to screen thermophiles for the production of bioactive metabolites, including unstable agents (Shih et al. 2014). In 2009, Yang et al. used ICD-MS and LC-SPE-NMR to identify the active components of a cytotoxic ethyl acetate extract of the thermophilic fungus *Malbranchea sulfurea* (Yang et al. 2009). Several halogenated metabolites were identified by ICD-MS in only the intact brown hyphae of *M. sulfurea*. LC-MS was used to screen crude extracts for these metabolites, and LC-SPE-NMR was used to elucidate the structures of the photosensitive polyketides malbranpyrroles A–F (**13–18**) (Fig. 2). The malbranpyrroles structurally differ based upon the presence of a chlorine atom as well as the substitution of the γ-pyrone ring and/or double bonds. Malbranpyrrole A (lacking a chlorine atom; **13**) did not display cytotoxic activity toward human pancreatic cancer PANC-1, liver cancer HepG2, and breast cancer MCF-7 cell lines, whereas malbranpyrroles C–F (**15–18**) did with IC_{50} values ranging from 3 to 11 μM. Not surprisingly, these results are consistent with halogens being atoms that improve the bioactivity of a molecule (Lu et al. 2012; Hernandes et al. 2010). Notably, in the presence of light, the malbranpyrroles isomerize; however, cytotoxic activity was not affected when cells were treated with malbranpyrroles in the presence of light. Based on the percentage of cells in the G_0/G_1 phase increasing in human breast cancer MCF-7 and liver cancer HepG2 cell lines, the malbranpyrroles are proposed to arrest the cell cycle at the G_0/G_1 phase. Naturally occurring polyvinyl pyrroles are very rare in nature, as only the rumbrins (Yamagishi et al. 1993), auxarconjuatins (Clark et al. 2006), wallemias (Ahmed et al. 1984), and keronopsins (Höfle et al. 1994) have been reported. Thus, the malbranpyrroles are a good example of structurally unprecedented, bioactive molecules that can be produced by thermophiles.

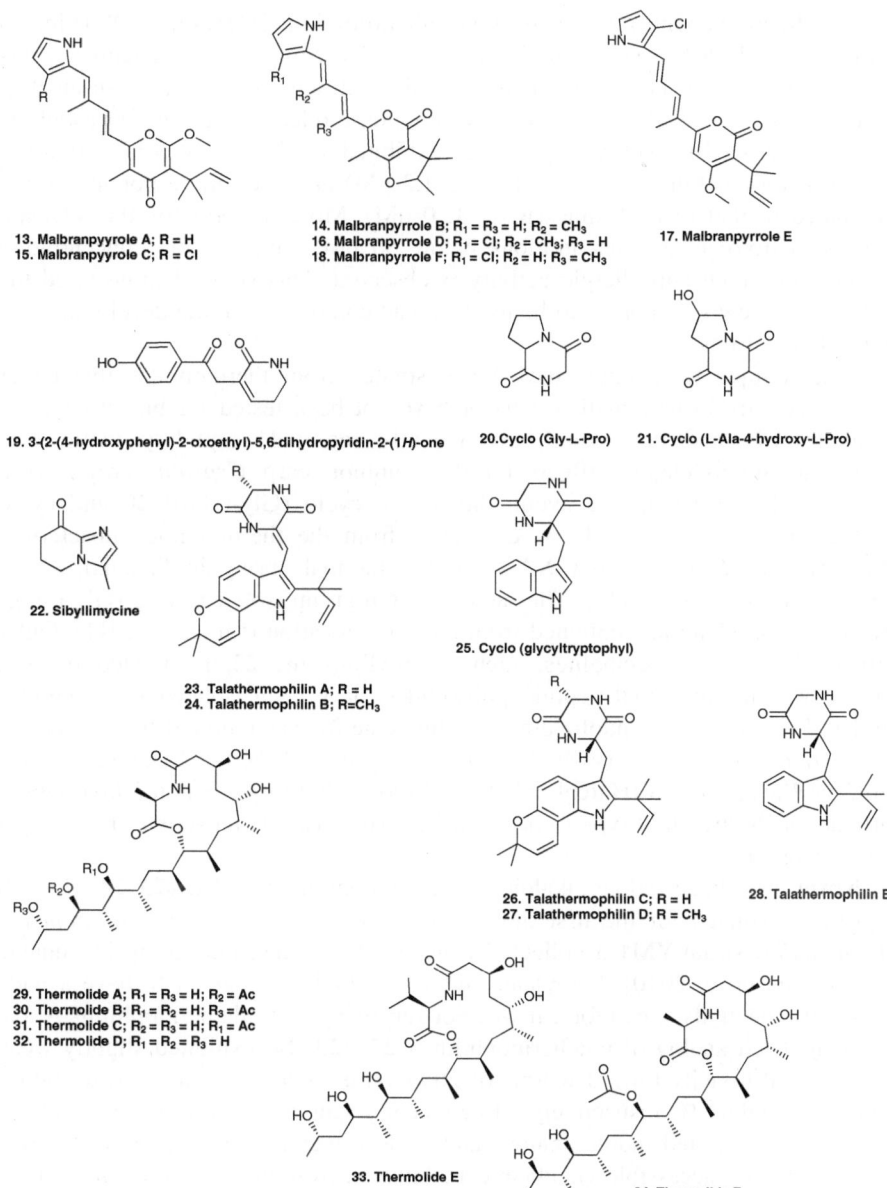

Fig. 2 Structures **13–34**

Computational chemistry has also been a useful tool for predicting the bioactivity of new metabolites from thermophiles. For example, the thermophilic soil fungus *Humicola grisea* var *thermoidea* was reported by Andriouli and coworkers to produce the anti-allergic δ-lactam derivative

(3-(2-(4-hydroxyphenyl)-2-oxoethyl)-5,6-dihydropyridin-2(1H)-one) **19** (Fig. 2) (Andrioli et al. 2012). Gratifyingly, using in silico computational chemistry, this δ-lactam was predicted and verified to exhibit anti-allergic activity with an IC_{50} value of 18.7 ± 6.7 µM using a β-hexosaminidase release assay in rat basophilic leukemia cells RBL-2H3. Compound **19** exhibited similar anti-allergic activity to that of ketotifen fumarate (IC_{50}, 15.0 ± 1.3 µM) and even more potent activity compared to that of azelastine (IC_{50}, 32.0 µM). More importantly, the δ-lactam was not cytotoxic at concentrations below 50 µM, which is the concentration window in which anti-allergic activity is observed. This study demonstrated the potential of new metabolites to be used as lead compounds in the development of anti-allergic drugs.

Other unique metabolites have been isolated from thermophiles that either lack bioactivity in mammalian cells or have not been tested for bioactivity. For example, Liu and coworkers have recently reported secondary metabolites with immunomodulatory effects on the common carp *Cyprinus carpio* (Liu et al. 2011). In 2010, two cyclic dipeptides, cyclo (Gly-L-Pro) **20** and cyclo (L-Ala-4-hydroxyl-L-Pro) **21,** were isolated from the thermophilic eubacterium *Anoxybacillus kamchatkensis* XA-1 from a thermal spring in Shaanxi, China (Fig. 2) (Wang et al. 2011a). The same research group isolated cyclo (Gly-L-Pro) **20** from *A. flavithermus* obtained from the same location (Liu et al. 2011). Other structurally unique metabolites, such as sibyllimycine **22**, prenylated indoles, and hybrid nonribosomal peptide–polyketides, have been isolated from thermophiles (Fig. 2). The azaindolizidine sibyllimycine **22** was reported from a species of *Thermoactinomycetes* isolated from a hot spring (60 °C) at Lake Tanganyika in Cape Banza, Africa (Hafenbradl et al. 1996). Sibyllimycine is the first natural product of its structural type to be published. However, its bioactivity has not yet been reported.

Novel prenylated indole alkaloids talathermophilins A and B (**23–24**) (Fig. 2) were also isolated for the first time from the thermophilic fungus *Talaromyces thermophiles* strain YM1-3 collected from the Tengchong hot spring in Yunnan, China (Chu et al. 2010). Prenylated indole alkaloids are known to have a wide range of bioactivity. Therefore, it was not surprising that talathermophilins A and B (methylated analog of talathermophilin A **23**) (**23–24**) exhibited slightly weak nematocidal toxicity toward worms of *Panagrellus redivevus* with growth inhibition of 38 and 44 %, respectively. More recently, Guo and coworkers also isolated three new prenylated indole analogs and cyclo (glycyltryptophyl) **25**, which was previously only accessible via chemical synthesis, from the *T. thermophiles* strain YM3-4 collected from the same location. Talathermophilins C–E (**26–28**) (Fig. 2) are tryptophan-derived indoles with that structurally differ by the presence of a methyl group or pyran ring (Guo et al. 2011). The bioactivity of these compounds was not reported. A year later, the same research group reported the isolation of an unprecedented class of a hybrid nonribosomal peptide–polyketides containing a 13-membered lactam-bearing macrolactone from the same fungal strain (Guo et al. 2012). Thermolides A–F (**29–34**) (Fig. 2) structurally differ by alkyl chain length, the presence of an acetonide unit, or acetylation, methylation, and hydroxylation

Table 1 Bioactive compounds isolated from terrestrial thermophiles

Thermophile	Optimal growth temperature (°C)	Source	Compound	Bioactivity	References
Thermoactinomyces antibioticus	50	Soil	Thermorubin **1**	Antimicrobial; aldose reductase inhibitor	Craveri et al. (1964), Johnson et al. (1980)
Myriococcum albomyces	47	Soil	Myriocin **2**	Antifungal; immunosuppressant	Craveri et al. (1972), Aragozzini et al. (1972), Bagli et al. (1973)
Streptomyces thermoviolaceus subsp. *pigens* var **WR-141**	*37	N/A	Dihydrogranaticin **3**	Antimicrobial	Pyrek et al. (1977), Brimble et al. (1999)
S. thermoviolaceus subsp. *pigens* var **WR-141**	*37	N/A	Bisanhydrogranaticin **4**	Antimicrobial	Pyrek et al. (1977)
S. thermoviolaceus subsp. *pigens* var **WR-141**	*37	N/A	Bisanhydrogranaticin alcohol derivative **5**	Antimicrobial	Pyrek et al. (1977)
Bacillus stearothermophilus UK563	60	N/A	Bis(2-hydroxyethyl)trisulfide (BS-1) **6**	Cytotoxic	Kohama et al. (1993)
P. akbaalia	50	Thermal springs	Pyochelin **7**	Siderophore; antifungal	Phoebe et al. (2001)
B. licheniformis VK21	45	Thermal springs	2,3-Dihydroxybenzoylglycyl-threonine (S$_{VK21}$) **8**	Siderophore; potential antimicrobial	Temirov et al. (2003), Esikova et al. (2002)

(continued)

Table 1 (continued)

Thermophile	Optimal growth temperature (°C)	Source	Compound	Bioactivity	References
Microbispora aerate	50	Penguin excrements	Microbiaeratin **9**	Weak antiproliferative/cytotoxic	Ivanova et al. (2007)
Thermobifida fusca	55	N/A	Fuscachelins A–C (**10–12**)	Siderophores	Dimise et al. (2008)
Malbranchea sulfurea	40	Soil	Malbranpyrroles A–C (**13–18**)	Cytotoxic	Yang et al. (2009)
Humicola grisea var *thermoidea*	40	Soil	(3-(2-(4-Hydroxyphenyl)-2-oxoethyl)-5,6-dihydropyridin-2(1*H*)-one) **19**	Anti-allergic	Andrioli et al. (2012)
Anoxybacillus kamchatkensis XA-1; *A. flavithermus*	60	Thermal spring	Cyclo (Gly-L-Pro) **20**	Immunomodulators of carp	Wang et al. (2011a), Liu et al. (2011)
A. kamchatkensis XA-1	60	Thermal spring	Cyclo (L-Ala-4-hydroxyl-L-pro) **21**	Immunomodulators of carp	Wang et al. (2011a)
Thermoactinomycetes sp.	60	Thermal spring	Sibyllimycine **22**	N/A	Hafenbradl et al. (1996)
Talaromyces thermophiles strain YM1-3	45	Thermal spring	Talathermophilins A–E (**23–24**; **26–28**)	Very weak nematocidals	Chu et al. (2010)
T. thermophiles strain YM3-4	45	Thermal spring	Cyclo (glycyltryptophyl) **25**	N/A	Chu et al. (2010)
T. thermophiles strain YM3-4	45	Thermal spring	Thermolides A–F (**29–34**)	Nematocidals	Guo et al. (2012)

N/A information not available

*Temperatures at which cultures were grown

patterns. Thermolides A and B (**29–30**) were determined to exhibit potent nematocidal activity with LD_{50} values ranging from 0.5 to 1 µg/ml. Remarkably, these reports demonstrate how different strains of *T. thermophiles* obtained from the same location have diverse metabolic profiles.

The thermophiles described demonstrate the biodiversity found in different thermal environments, even among different strains of the same species. rRNA and metagenomic sequence analyses (Barns et al. 1994; Pride and Schoenfeld 2008; Inskeep et al. 2010) have revealed the scores of unidentified, diverse geothermal microbes (Rademacher et al. 2012; Kröber et al. 2009). Table 1 summarizes all of the secondary metabolites described in this chapter and their known bioactivity. Over time, new pharmacophores will be found with an increase in the number of screening programs and metagenomic sequencing of thermophiles, especially from the least evolved microorganisms among eubacteria, eukaryotes, and archaea, such as anaerobic hyperthermophilic archaea and eubacteria, which represent the most extreme and divergent thermophiles (Stetter 1996).

3 Psychrophiles

Morita used the term "psychrophiles" to refer to microorganisms that grow optimally at (extremely) cold temperatures ranging from 15 °C or below (e.g., −20 °C in sea ice) to a maximum temperature of 20 °C (Morita 1975). Roughly 80 % of our planet's biosphere is at temperatures below 5 °C, which includes most of the world's oceans (70 % of the Earth's surface), sea ice, Antarctica, parts of North America and Europe, the deep-sea, mountain regions, as well as the mesosphere and stratosphere (de Maayer et al. 2014). Thus, most microorganisms on Earth can be considered to be psychrophiles. While various microorganisms are preserved at −196 °C in liquid nitrogen, the lowest recorded temperature for active microbial growth is −18 °C, which is the temperature at which *Rhodotulura glutinis* causes food spoilage (Schroeter et al. 1994). To survive at such low temperatures, psychrophiles have evolved ways to overcome several challenges, which include decreased enzyme activity, several lethal freeze–thaw cycles, improper protein folding, decreased membrane fluidity, decreased rates of transcription, translation, and cell division, as well as altered transport of nutrients and waste (D'Amico et al. 2006). Psychrotolerant microorganisms grow and reproduce optimally at almost mesophilic temperatures between 20 and 40 °C and commonly inhabit cold environments because they have better nutritional adaptability (Wynn-Williams 1990) or have undergone horizontal gene transfer from mesophiles (Aislabie et al. 2004).

Foster published the first report of eubacteria growing at 0 °C or below over a century ago in 1887 (Foster 1887). Then, in 1918, McClean identified Gram-positive cocci as well as Gram-negative spore- and nonspore-forming rods (McClean 1918). Over 30 years later, Darling and Siple isolated 178 microbial strains from snow, ice, and frozen algae obtained from Antarctica (Darling and Siple 1941). Since then, psychrophiles have only been extensively studied within the last 30–40 years. Some psychrophiles are considered to be halophile and/or oligotrophic, microorganisms with limited access to nutrient resources, due to slower diffusion rates of dissolved gases and inorganic nutrients. Psychrophiles overcome these challenges by undergoing seasonal dormancy, incorporating more fatty acids (or antifreeze molecules) to maintain membrane fluidity and the transport of nutrients and waste (Russell 1989), as well as expressing highly specific enzymes at low temperatures and producing cryoprotectants to prevent cell lysis during freeze–thaw cycles. Because the solubility of gases increases due to retarded diffusion rates and the production of reactive oxygen species increases, diverse chemical structures of bioactive secondary metabolites are expected.

Commercial manufacturers are interested in these cold-loving organisms for the bioremediation of polar soils and as sources of enzymes, such as β-glucanases, cellulases, lipases, pectinases, and proteinases, which are active at refrigerator temperatures. These enzymes have been used on industrial scales to produce detergents, food additives, and biosensors. However, over the past 20 years, an increasing number of reports have been published on psychrophiles producing bioactive metabolites, especially in marine environments, making these microbes an unexplored source of new pharmacophores. Several screens to isolate new bioactive secondary metabolites have been conducted on organic extracts of psychrophiles within the past decade. For the purpose of this book, the present chapter focuses only on metabolites isolated from cold terrestrial climates.

The number of publications on the identification of novel molecules produced by psychrotolerant microbes has significantly increased between the years of 2003 and 2013. In 2004, the psychrotolerant fungus Penicillium reibeum (IBT 16537) was isolated from soil under a redcurrant bush in the tundra of Wyoming, USA, and reported to produce two new fungal metabolites, psychrophilin A **35** and cycloaspeptide D **36** (Fig. 3) (Dalsgaard et al. 2004b; Frisvad et al. 2006). Psychrophilin A **35** is composed of three amino acids and the first cyclic peptide from a natural source to contain a nitro group instead of an amino group. The pentapeptide cycloaspeptide D **36** has a similar structure to that of cycloaspeptide A but possesses a valine residue instead of a leucine residue.

Dalsgaard and coworkers later isolated several new psychrophilins, some of which were determined to have novel bioactivity (Dalsgaard et al. 2004a). Psychrophilins B and C (**37–38**) (Fig. 3) were isolated from another psychrotolerant fungus, *P. rivulum* Frisvad, an isolate maintained at the IBT Culture Collection at BioCentrum-DTU, Technical University of Denmark (Dalsgaard et al. 2004a). No bioactivity has been reported for these compounds. However, this fungus was later reported to produce the structurally complex alkaloids communesins G and H (**39–40**) (Fig. 3) (Dalsgaard et al. 2005a), which are polycyclic tryptamine molecules containing an epoxide moiety with varying amide alkyl chain lengths. Interestingly, unlike other communesins isolated from marine-derived fungi (Jadulco et al. 2003; Numata et al. 1993), compounds **39** and **40** did not exhibit any cytotoxic, antimicrobial, or antiviral activity (Numata et al. 1993; Jadulco et al. 2003).

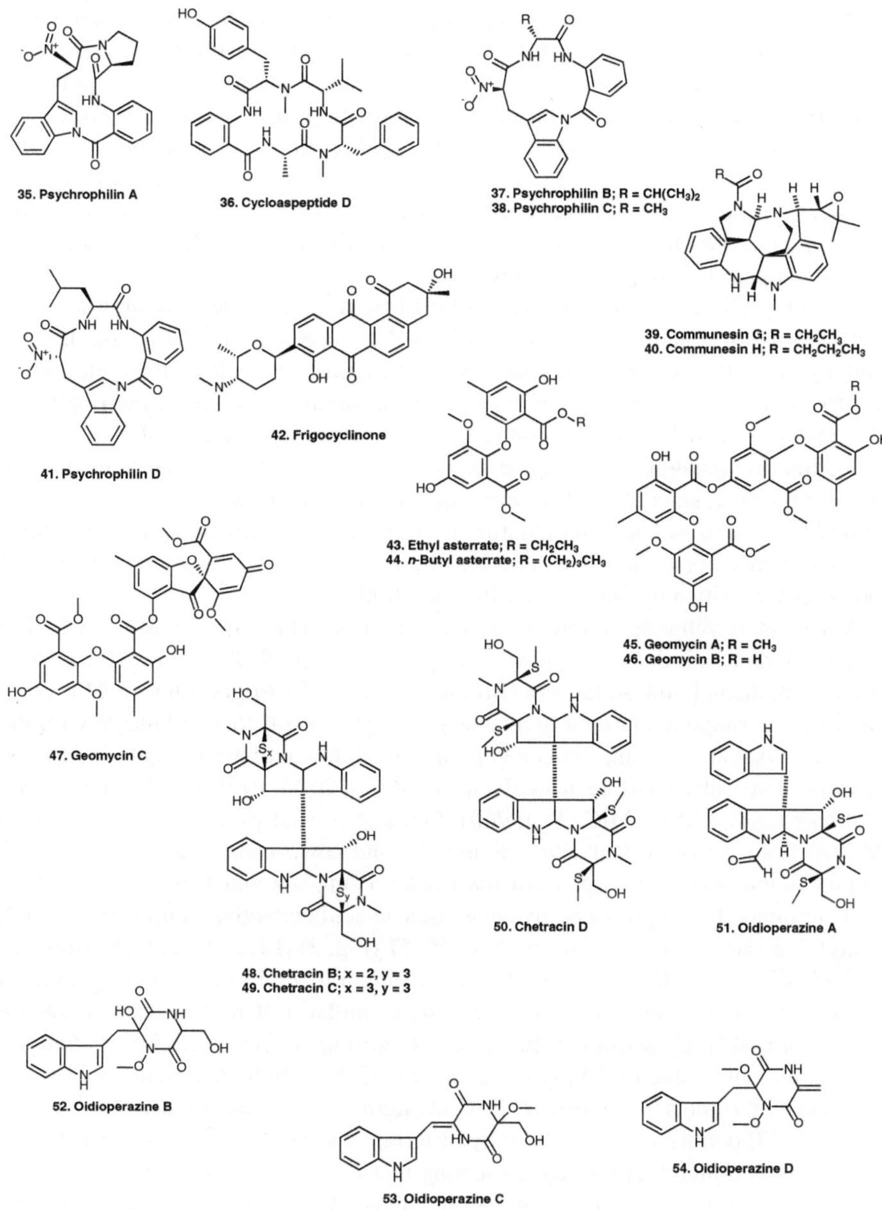

Fig. 3   Structures 35–54

Psychrophilin D **41** (Fig. 3) was later reported to be produced by a psychrotolerant fungus, *P. algidium* (IBT 22067), isolated from soil under a *Ribes* sp. East of Oksestien, Zackenberg, Greenland (Dalsgaard et al. 2005b). Psychrophilin D **41** exhibited cytotoxic activity against murine leukemia P-388 cells with an IC_{50} value of 23 μM. Furthermore, cycloaspeptide D **36** exhibited moderate antiplasmodial activity with an IC_{50} value of 7.5 μM. Interestingly, a psychrophilin possessing an acetylated α-amino group in the tryptophan residue (psychrophilin E) was recently isolated from the cocultivation of two marine alga-derived fungi of the genus *Aspergillus* (Ebada et al. 2014); however, this compound did not exhibit cytotoxic activity. Nevertheless, these studies demonstrate the diverse chemical entities and biological activity that can be found in different locations in extremely cold climates. Moreover, slight modifications to the core molecular structure appear to reduce cytotoxicity.

In 1999, Nichols and coworkers published a review describing how relatively few microorganisms had been collected from Antarctica due to the lack of funding for collecting and maintaining psychrophiles as well as their identification (Nichols et al. 1999). In addition, the biodiversity found in psychrophiles at this time was also limited, as public culture collections, such as the Australian Collection of Antarctic Microorganisms (ACAM), only consisted of known strains. However, since the 1990s, more new microbial strains from Antarctica and beyond have been isolated and identified. For a current summary of psychrophiles isolated from diverse environments, see the review by Buzzini et al. (2012) and book chapter written by Hoover and Pikuta (2009).

A new angucyclinone antibiotic, frigocyclinone **42** (Fig. 3), was isolated from a psychrotolerant, soil-derived *Streptomyces griseus* strain NTK 97 from Terra Nova Bay at Edmunson Point, Antarctica (Bruntner et al. 2005). Frigocyclinone **42** is composed of a tetrangomycin moiety attached through a *C*-glycosidic linkage with the aminodeoxysugar ossamine. This compound exhibited antibacterial activity against Gram-positive eubacteria, such as *Bacillus subtilis* DSM 10 (MIC, 4.6 μg/ml) and *S. aureus* DSM 21231 (MIC, 15 μg/ml). Other terrestrial psychrophiles have been reported from Antarctic soil. For example, Li and coworkers isolated the psychrotolerant fungus *Geomyces* sp. from the Foldes Peninsula and King George Island and determined that it produced five new asterric acid derivatives, ethyl asterrate **43**, *n*-butyl asterrate **44**, and geomycins A–C **45–47** (Fig. 3) (Li et al. 2008). Geomycins A–C **45–47** structurally differ based upon lacking a methyl ester moiety (geomycin B **46**) and/or possessing a spirotricyclic moiety similar to that of bisdechlorogeodin (geomycin C **47**). Geomycin B **46** exhibited antifungal activity against *A. fumigatus* with an MIC value of 20 μg/ml. Geomycin C **47** exhibited antimicrobial activity against Gram-positive *S. aureus* (MIC, 24 μg/ml) and Gram-negative *Escherichia coli* (MIC, 20 μg/ml), respectively. Ethyl asterrate **43** and *n*-butyl asterate **44** did not exhibit any antimicrobial activity, indicating that ester chain length does not play a role in the antimicrobial activity of asterric acids. We speculate that the carboxylic acid moiety in geomycin B **46** is important for antifungal activity, making it the first asterric acid derivative reported to have this bioactivity.

Recently, Li and coworkers isolated two new epipolythiodioxopiperazines, chetracins B and C (**48–49**), and five new diketopiperazines, chetracin D **50** and oidioperazines A–D (**51–54**), from the psychrophilic fungus *Oidiodendron truncatum* GW3-13 collected from the soil under lichens in Antarctica (Fig. 3) (Li et al. 2012). Oidioperazines A–D (**51–54**) appear to be intermediates in the biosynthesis of the chetracins. Chetracins B and C (**48–49**) contain different numbers of sulfur atoms within a bridged polysulfide piperazine ring, which has been found in metabolites with cytotoxic, immunomodulatory, antiviral, antimicrobial, and antiproliferative activity. Chetracin D **50** is simply the reduced form of the known diketopiperazine melinacidin IV. Not surprisingly, nanomolar concentrations of the polysulfide chetracin B **48** exhibited potent cytotoxic effects against human colorectal cancer HCT-8, hepatocellular adenocarcinoma Bel-7402, gastric cancer BGS-823, lung adenocarcinoma A549, and ovarian cancer A2780 cell lines with ranging IC_{50} values from 3 to 28 nM. Potent activity was also exhibited by melinacidin IV, which has one less sulfur atom in one of its oxopiperazine rings compared to compound **50**. Chetracins C and D (**49–50**) were also reported to exhibit significant cytotoxic activity at low micromolar IC_{50} values, ranging from 0.14 to 1.65 µM, with chetracin C **49** being the most potent, suggesting that the polysulfide ring does enhance the cytotoxicity of the chetracins.

In addition to producing structurally complex secondary metabolites, psychrophiles isolated from very extreme environments, such as ice, have also been reported to incorporate unique functional groups in molecules. An interesting psychrotolerant Arctic Sea eubacterium, *Salengentibacter* sp. T436, was isolated from the bottom section of a sea floe by Al-Zereini et al. (2007) and reported to produce 19 aromatic nitro compounds, including four new metabolites (**55–58**) and six previously reported metabolites (**59–64**) that were isolated from a natural source for the first time (Fig. 4). None of the newly isolated metabolites exhibited antimicrobial activity, but some displayed phytotoxic activity. We are not exactly sure of the depth at which *Salengentibacter* sp. T436 dwells (i.e., in the ice or in the water underneath the ice); however, microorganisms have been isolated from the liquid veins of ice concentrated with both nitric acid and sulfuric acid (Price 2000), which may be involved in installing nitro functional groups into their corresponding secondary metabolites. Additionally, water below ice is enriched with oxygen and other gases, as diffusion rates are retarded, which could also be involved in the formation of nitro groups. These microbes may also use a slow process utilizing their own limited resources to produce nitro compounds. In any case, this is an excellent example of how resources within a certain environment could result in unique metabolic profiles.

Other screening programs of psychrophiles have been initiated to isolate new metabolites with unverified structures or bioactivity as well as structurally interesting metabolites that lack bioactivity. For example, Mojib and coworkers have searched for bioactive eubacterial pigments produced by psychrophiles and isolated violacein 6 (J-PVP; **65**) and flexirubin 7 (F-YOP; **66**) from *Janthinobacterium* sp. Ant5-2 and *Flavobacterium* sp. Ant342, respectively (Fig. 4) (Mojib et al. 2010). These eubacterial strains were isolated from landlocked freshwater lakes in Schirmacher Oasis, East Antarctica. However, the structures of J-PVP **65** and F-YOP **66** have not been fully elucidated. J-PVP **65** is a violet pigment that exhibits antimycobacterial activity against *Mycobacterium smegmatis* mc^{2}155 (MIC, 8.6 µg/ml), *M. tuberculosis* mc^{2}6230 (MIC, 5 µg/ml), and virulent *M. tuberculosis* (MIC, 34.4 µg/ml). F-YOP is a yellow orange pigment that also exhibits antimycobacterial activity against *M. smegmatis* mc^{2}155 (MIC, 3.6 µg/ml),

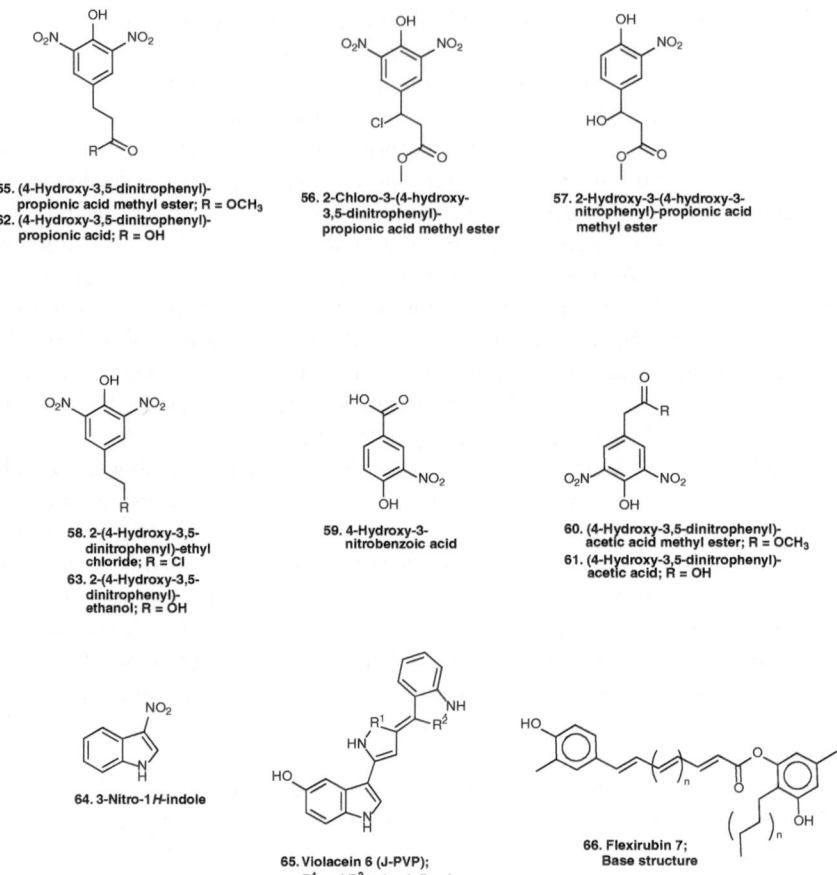

Fig. 4 Structures **55–66**

$M.$ $tuberculosis$ mc^{2}6230 (MIC, 2.6 μg/ml), and virulent $M.$ $tuberculosis$ (MIC, 10.8 μg/ml). Pigment production in psychrophiles is not surprising, as nonpolar carotenoid pigments are commonly used by these microorganisms to adapt to cold environments (D'Amico et al. 2006). Once their structures have been fully elucidated, J-PVP **65** and F-YOP **66** may serve as lead compounds for the development of antimycobacterial drugs for the treatment of tuberculosis. Recently, the gene cluster involved in polyene biosynthesis en route to producing flexirubin in the eubacterium $Chitinophaga$ $pinesis$ was reported, suggesting that other derivatives may be obtained via genetic engineering and these genes could serve as a metabolic marker to identify other microbial sources of bioactive flexirubin-type compounds (Schöner et al. 2014).

Shekh and coworkers conducted another screen for bioactive compounds produced by psychrophiles. Antifungal activity was detected against the

Table 2 Bioactive compounds isolated from terrestrial psychrophile

Psychrophile	Optimal growth temperature (°C)	Source	Compound	Bioactivity	References
Penicillium reibeum (IBT 16537) (**35**); *P. rivulum* Frisvad (**37–38**); *P. algidium* (IBT 22067) (**41**)	25	Soil	Psychrophilins A–D **35**, **37–38, 41**	Antifungal (**35**); cytotoxic (**41**);	Dalsgaard et al. (2004)
P. reibeum (IBT 16537)	25	Soil	Cycloaspeptide D **36**	Antifungal; antiplasmodial activity	Dalsgaard et al. (2004)
P. rivulum Frisvad	20	N/A	Communesins G and H (**39–40**)	Antimicrobial	Dalsgaard et al. (2005)
Streptomyces griseus strain NTK 97	27	Soil	Frigocyclinone **42**	Antimicrobial	Bruntner et al. (2005)
S. thermoviolaceus subsp. *pigens* var WR-141	25	Soil	Ethyl asterrate **43**	N/A	Li et al. (2008)
Bacillus stearothermophilus UK563	25	Soil	*n*-Butyl asterrate **44**	N/A	Li et al. (2008)
Geomyces sp.	25	Soil	Geomycins A–C **45-47**	Antimicrobial	Li et al. (2008)
Oidiodendron truncatum GW3-13	15	Soil	Chetracins B and C (**48–49**); chetracin D **50**	Cytotoxic	Li et al. (2012)
O. truncatum GW3-13	15	Soil	Oidioperazines A–D (**51–54**)	N/A	Li et al. (2012)
Salengentibacter sp. T436	21	Underneath sea floe	19 aromatic nitro compounds (**55–64**)	Phytotoxic	Al-Zereini et al. (2007)
Janthinobacterium sp. Ant5-2	*15	Freshwater lake	Violacein 6 (J-PVP; **65**)	Antimycobacterial	Mojib et al. (2010)
Flavobacterium sp. An-342	*15	Freshwater lake	Flexirubin 7 (F-YOP; **66**)	Antimycobacterial	Mojib et al. (2010)

N/A information not available

* Temperature was assumed based on another publication using the same stain by the authors (Huang et al. 2012a)

multidrug-resistant pathogenic yeast *Candida albicans* NCIM 3471 from seven strains of psychrotolerant eubacteria isolated from the following locations: lake sediments; water around glacier mouths, streams, sea convergence, and permafrost soils; and feces, feathers, and soils collected from penguin rookeries in the Arctic and Antarctica (Shekh et al. 2011). An isolate from the penguin rookery was determined to be *Enterococcus faecium*, which displayed potent antifungal activity. The authors subsequently reported the active component to be a 40-amino acid antifungal peptide, which was confirmed by the direct detection by polyacrylamide gel electrophoresis. These screening programs further highlight the unlimited potential of psychrophiles to produce unique bioactive agents, including pigments.

Table 2 summarizes the natural products produced by the psychrophiles or psychrotolerant microorganisms discussed in this chapter as well as the temperatures used to cultivate them and any reported bioactivity. Although the number of bioactive secondary metabolites from nonmarine (terrestrial, excrements, ice, or fresh water) sources is limited, we believe more metabolites will be reported with the increase in the isolation and cultivation of new psychrophiles. Scientists have only recently realized the remarkable biodiversity that exists among psychrophiles, which is reflected by the variation in the microbial communities in the soil habitats of Antarctica influenced by the glacial history of the continent (Cowan et al. 2014). Metagenomic sequencing has also revealed the genome plasticity of psychrophiles in response to adapting to the cold and the different survival strategies used by each microorganism (de Maayer et al. 2014). Furthermore, with increasing global warming, microbial communities growing in cold climates are predicted to change significantly along with the availability of nutrients and defense mechanisms for survival. Thus, we expect to see even more diverse metabolic profiles with potentially new pharmacophores.

4 Acidophiles

Acidic environments form as a result of geochemical activities and/or anthropogenic influences, such as deep and surface coal/metal mining (e.g., acid mine drainage water or mine spoils). Microorganisms that grow optimally in highly acidic environments are called acidophiles. Extreme acidophiles typically dwell in environments with a pH value <3, and moderate acidophiles grow optimally in habitats with a pH value between 3 and 5. Most acidic environments on Earth result from the biologically accelerated oxidation of elemental sulfur and sulfide minerals by molecular oxygen or ferric iron, producing enough sulfuric acid to lower the pH of terrestrial and marine environments. Not surprisingly, many acidophiles are resistant to metal-rich environments (i.e., metalloresistant) and can generate energy or ATP from metals, such as ferrous iron. The first reported acidophile was a sulfur-oxidizing bacterium, *Acidothiobacillus thiooxidans* (formerly known as *Thiobacillus thiooxidans*), isolated from soil–rock–sulfur composts in 1922 by Waksman and Joffe (1922). Acidic niches, such as mud pots, acid springs, acidic mine drainage, or hydrothermal vents, are commonly places with high concentrations of sulfidic minerals as well as a wide range of temperatures and redox potentials. Examples of these environments are the solfatara fields around Yellowstone National Park, USA; Wyoming, USA; Whakerewa, New Zealand; and Krisuvik, Iceland.

The primordial conditions on Earth are thought to be acidic, sulfurous, high in temperature, and volcanic, suggesting that acidophiles are relics of some of the world's first microorganisms (Johnson 2007). Thus, from an evolutionary standpoint, the physiology and ecology of acidophiles are quite remarkable, as their genetic code (Wächtershäuser 2006) and metabolic processes most likely evolved from acidic environments (e.g., on sulfide minerals). To survive in low pH environments, acidophiles excrete acid outside of the cell to maintain a pH gradient across the plasma membrane of several pH units for cellular function, ensuring that biological reactions can occur within a pH range of 5.0–7.5. These microorganisms, commonly archaea and eubacteria, have reverse membrane potentials (negative intracellular membrane potentials), a large number of secondary transporters, extracellular enzymes that can function at extremely low pH, and membranes that are highly impermeable to H_3O^+. Acidophiles are also widely distributed among *Eukarya* but limited to species that have developed other strategies to tolerate dissolved metal and increased H_3O^+ concentrations.

Acidophiles are sources of acid-stable enzymes extensively used in bioremediation as well as excellent models for understanding the genetic and biochemical basis of pH homeostasis. Although there are few reports of acidophiles producing new bioactive metabolites, these microorganisms undoubtedly have the potential to yield unusual pharmacophores. Seminal studies have been published on isolating bioactive molecules via bioprospecting the Berkeley Pit in the mining mecca of Butte, Montana, USA. The Berkeley Pit is part of the largest Environmental Protection Agency (EPA) Superfund site in North America. The pit is an abandoned, 1.6-km-wide, and 457-m-deep pit gorged out of the Boulder Batholith located at the head of the Columbia River ecosystem and surrounded by several deep mining shafts (Stierle and Stierle 2005). After mining ceased in 1982, the underground tunnels were no longer dewatered and groundwater began to accumulate in the pit. Now three decades later, 150 billion liters of water is contained within the limestone-based pit, which is filled with 15 million liters of water daily. Iron pyrite and other metal sulfates (e.g., 1,200 ppm iron, 240 ppm copper, 290 ppm aluminum, and 650 ppm zinc) in the surrounding area significantly influence the nature of the Berkeley Pit. Iron pyrite is quickly oxidized by molecular oxygen or ferric iron dissolved in acidic water to produce sulfuric acid, which dissolves minerals and ultimately lowers the pH of the pit. Based on the possible oxidation reactions that can take place, the pH of the Berkeley Pit is within the range of 2.5–2.7. Under these conditions, the main acidophilic eubacterium present is *Acidithiobacillus ferrooxidans*, which survives by producing ATP via increasing the rate at which ferrous iron is oxidized by molecular oxygen (Stierle and Stierle 2005).

In 1995, Steirle and coworkers began bioprospecting the Berkeley Pit because it was a new and exotic ecosystem. Ten years later, 60 fungi and eubacteria were isolated from the surface to lake bottom sediments (Stierle and Stierle 2005).

Initially, Stierle and coworkers screened organic extracts of species of *Penicillium* or *P. rubrum* isolated from a depth of 270 m for secondary metabolites with activity against *S. aureus* and brine shrimp lethality (Stierle et al. 2004). Two in vitro signal transduction bioassays were also used to find new inhibitors of metalloproteinase-3 (MMP-3) and caspase-1 (casp-1) enzymes, active proteases with a promising correlation to the NCI-Developmental Therapeutic Program (NCI-DTP) human cancer cell line assay. Using this screen, several new structural classes of bioactive secondary metabolites were reported from either a single isolate of *P. rubrum or* species of *Penicillium*, including berkeleydione **67**, berkeleytrione **68**, berkeleyones (**69–71**), berkelic acid **72**, berkeleyacetals (**73–75**), berkeleyamides (**76–79**), berkazaphilones (**80–81; 84**), and analogs.

4.1 Berkeleydione, Berkeleytrione, and Berkeleyones A–C

Berkeleydione **67** and berkeleytrione **68** (Fig. 5), hybrid polyketide–terpenoids that structurally differ in their A and B rings, were isolated from *P. rubrum* collected at a depth of 270 m (Stierle et al. 2004). Metabolites from this sesquiterpene structural class have also been isolated from plants, a basidiomycete, octocoral, sponges, red algae, as well as two fungal strains of *Fusarium sambucinium* and *P. roqueforti* (Stierle and Stierle 2013). Berkeleydione **67** and berkeleytrione **68** both inhibited MMP-3 and casp-1 at micromolar levels. Notably, berkeleytrione **68** exhibited selective activity against non-small cell lung cancer NCI-460 cell lines with a GI_{50} value of 0.39 μM. That same year, three structurally related analogs, berkeleyones A–C (**69–71**), were reported to be produced by this fungus (Stierle et al. 2011). Berkeleyones A–C (**69–71**) possess identical C and D rings to those in berkeleydione **67** and berkeleytrione **68** but possess A and B rings of different sizes as well as oxidation and methylation patterns. The berkeleyones inhibited the production of the interleukin-1β (IL-1β) cytokine, which is activated by casp-1, by inflammasomes in induced THP-1 cell line (a pro-monocytic leukemia cell line) assays with IC_{50} values ranging from 2.7 to 34.3 μM. In addition, berkeleyone B **70** completely inhibited casp-1 activity (100 % inhibition) similar to the positive control Ac-VYAD-CHO. Docking studies have been performed with casp-1 and either berkeleydione **67**, berkeleytrione **68**, or berkeleyones A–C (**69–71**), and the results revealed that these compounds bind to the active site cleft of the protease.

4.2 Berkelic Acid

In 2006, a *Chlorella mutabilis*-associated *Penicillium* sp. from the Berkeley Pit was reported by Stierle et al. to produce berkelic acid **72**, a unique spiroketal (Fig. 5) (Stierle et al. 2006). Berkelic acid **72**, belonging to a rare group of chroman spiroketals, inhibited both MMP-3 and casp-1 enzymes with GI_{50} values of

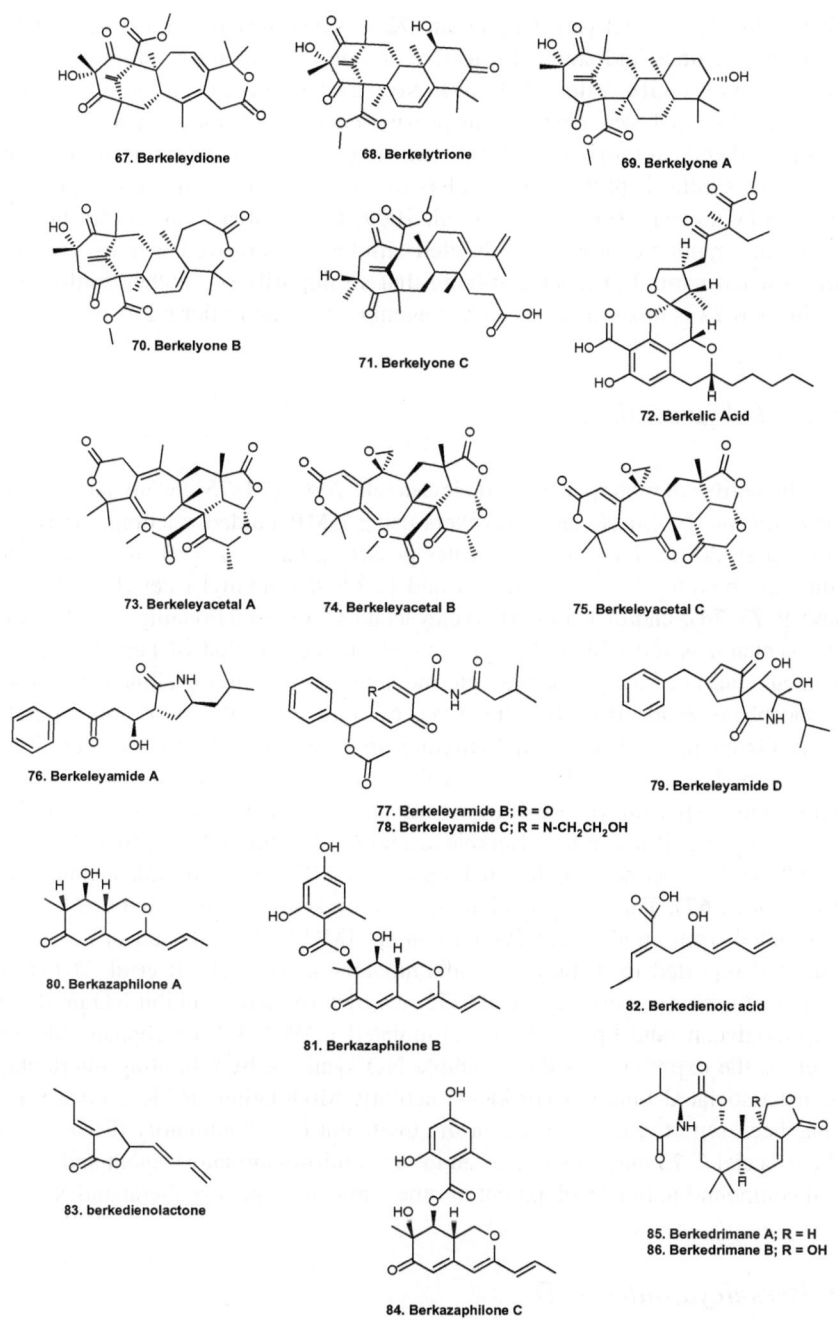

67. Berkeleydione

68. Berkelytrione

69. Berkelyone A

70. Berkelyone B

71. Berkelyone C

72. Berkelic Acid

73. Berkeleyacetal A

74. Berkeleyacetal B

75. Berkeleyacetal C

76. Berkeleyamide A

77. Berkeleyamide B; R = O
78. Berkeleyamide C; R = N-CH₂CH₂OH

79. Berkeleyamide D

80. Berkazaphilone A

81. Berkazaphilone B

82. Berkedienoic acid

83. berkedienolactone

84. Berkazaphilone C
(Sch 725680)

85. Berkedrimane A; R = H
86. Berkedrimane B; R = OH

Fig. 5   Structures **67–86**

1.87 and 98 µM, respectively. Compound **72** was also tested against the NCI-DTP 60 cancer cell lines and showed selective activity toward the ovarian cancer line OVCAR-3 with a GI_{50} value of 91 nM. No correlation between the up- or down-regulation of casp-1 was observed in ovarian cancer cell lines; however, MMP-3 was upregulated in only OVCAR-3 cell lines. Several research groups have reported the synthesis of this spiroketal as well as derivatives, including epi-22 ber-kelic acid (Wu et al. 2009; Bender et al. 2009; Buchgraber et al. 2008); however, none of the synthetic products exhibited similar or improved activity compared to the isolated natural product, implying that an impurity or another conformer of berkelic acid **72** is most likely the active agent (Stierle and Stierle 2013).

4.3 Berkeleyacetals A–C

Three bioactive monoterpenes, berkeleyacetals A–C (**73–75**), were later isolated from *P. rubrum* by Stierle and coworkers using NMR-guided fractionation (Fig. 5) (Stierle et al. 2007). These monoterpenes possess a rare 6–7 A–B ring system con-taining an epoxide (berkeleyacetals B and C **74–75**), methyl ester (berkeleycetals A and B **73–74**), and/or ketone (berkeleyacetal C **75**) functionality. The berkeley-acetal skeleton is most likely biosynthetically related to that of berkeleydione **67** via the alkylation of the polyketide intermediate 3,5-dimethylorsellinate by farnesyl pyrophosphate. Aside from berkeleydione **67**, only two other molecules, paraher-quonin (Okuyama et al. 1983) and citreonigrin A (Rusman 2006), have a 6–7 A–B ring system. Berkeleyacetal C **75** was the only compound determined to inhibit MMP-3 and casp-1 enzymes with micromolar IC_{50} values and tested in the NCI-DTP 60 cancer cell line assay. Berkeleyacetal C **75** inhibited the growth of leuke-mia cells as well as non-small cell lung cancer NCI 460 cells (also inhibited by berkeleydione **67**). This compound along with berkeleyacetals B and C **74–75** was also isolated from another soil *Penicillium* sp. IMU-0035 collected in Fukushima, Japan, and reported to display anti-inflammatory activity (Etoh et al. 2013). The most potent inhibitor, berkeleyacetal C **75**, was determined to inhibit NO production in peptidoglycan- and liposaccharide-stimulated RAW264.7 macrophage-like cells as well as the expression of the inducible NO synthase by inhibiting interleukin-1 receptor-associated kinase-4 (IRAK-4) activity. Modulating IRAK-4 kinase activ-ity has been reported to be useful in the treatment of inflammatory diseases; thus, berkeleyacetal C **75** may function as an in vivo anti-inflammatory agent and serve as a lead compound in the development of other anti-inflammatory therapeutics.

4.4 Berkeleyamides A–D

Berkeleyamides A–D (**76–79**) (Fig. 5) were also isolated from *P. rubrum* by the Stierle research group based on their inhibitory activity against MMP-3 and casp-1 proteases (Stierle et al. 2008). These compounds all possess terminal ben-zyl and isobutyl moieties but have a disubstituted γ-lactam (berkeleyamide A **76**), disubstituted 4-pyrone ring (berkeleyamide B **77**), disubstituted 4-pyridone ring (berkeleyamide C **78**), or azaspirocyclic ring system (berkeleyamide D **79**). Berkeleyamides A–C (**76–78**) have also been speculated to be on-pathway biosynthetic intermediates to form berkeleyamide D **79**. Furthermore, the origin of C19 is quite interesting as the carbonyl group may be derived from a leucine residue via a cyclic intermediate, demonstrating the intricate and complex chemistry performed by Nature. The total synthesis of (-)-berkeleyamide A performed by the Brimble research group confirmed that berkeleyamides are in fact derived from L-leucine. All compounds inhibited MMP-3 and casp-1 with low micromolar IC_{50} values, but berkeleyamides A **76** and D **79** were more potent inhibitors with IC_{50} values of 0.33 and 0.61 μM, respectively. This activity may be due to the presence of the almost terminal γ-lactam or azaspirocyclic ring. However, when these compounds were tested against NCI-DTP 60 cancer cell lines, they did not meet the criteria to undergo further testing.

4.5 Berkazaphilones A–C

The new berkazaphilones A and B (**80–81**), octadienoic acid derivatives berkedienoic acid **82** and berkedienolactone **83**, as well as other known metabolites, including azaphilone Sch 725680, have also been isolated from *P. rubrum* (Fig. 5) (Stierle et al. 2012a). Berkazaphilone A **80** is an α,β,γ,δ-unsaturated ketone bicyclic ring system, and berkazaphilone B **81** is its hydroxy derivative. Berkazaphilone C **84** may have been previously reported to be azaphilone Sch 725680 (Yang et al. 2006), an isomer of berkazaphilone B **81**; however, Stierle and coworkers were not sure if these compounds were indeed the same due to the lack of optical rotation data available for azaphilone Sch 725680. Berkazaphilones B **81** and C **84** exhibited potent inhibition against casp-1 both with an IC_{100} value of 25 μM, whereas berkazaphilone A **80** completely inhibited casp-1 at 250 μM. In addition, berkazaphilones B **81** and C **84** completely inhibited the production of IL-1β in THP-1 cells at concentrations of 5 and 50 μM, respectively. When compounds **80–81** and **84** were tested in the NCI-DTP 60 cancer cell line assay, both compounds selectively inhibited the growth of leukemia cell lines. Berkazaphilone B **81** exhibited a GI_{50} value of 2.1 μM against the RPMI-8226 cancer cell line and berkazaphilone C **84** exhibited a GI_{50} of 0.38 μM against the SR cell line. Interestingly, casp-1 was upregulated in almost every leukemia cell line and these compounds were active against. The other new metabolites, berkedienoic acid **82** and berkedienolactone **83**, appear to be biosynthetically related, such that berkedienolactone **83** is derived from the ring closure of berkedienoic acid **82**. However, these compounds have not been tested for bioactivity due to their impurities.

4.6 Other Metabolites

Other *Penicillium* species were isolated from the Berkeley Pit and reported to produce unique bioactive compounds. The α-β-unsaturated γ-butyrolactones

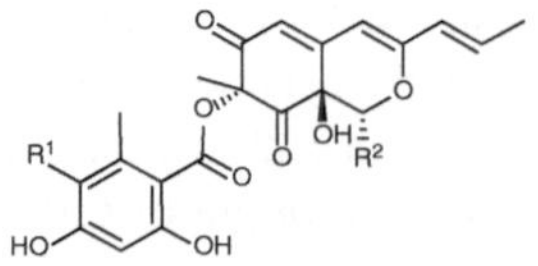

87. Purpurquinone A; R^1 = H, R^2 = OH
88. Purpurquinone B; R^1 = OH, R^2 = OH
89. Purpurquinone C; R^1 = H, R^2 = H

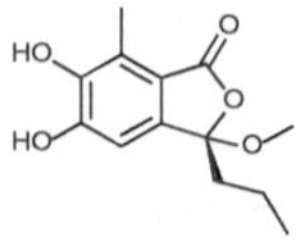

90. Purpurester A

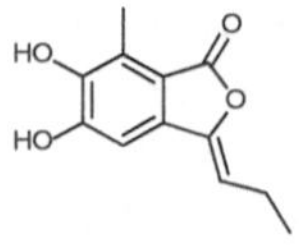

91. Purpurester B

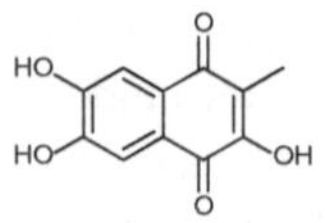

92. 2,6,7-Trihydroxy-3-methylnaphthalene-1,4-dione

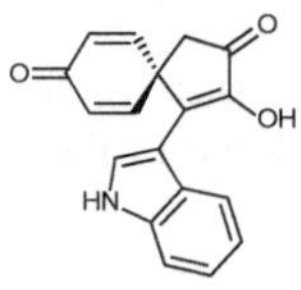

93. Spirobacillene A

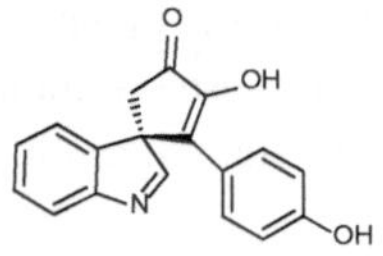

94. Spirobacillene B

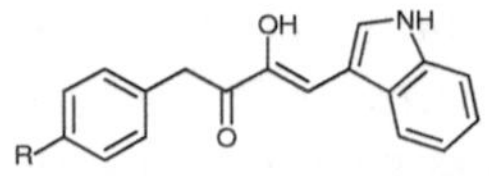

95. (Z)-3-hydroxy-4-(3-indolyl)-1-hydroxyphenyl-2-butenone; R = OH
96. (Z)-3-hydroxy-4-(3-indolyl)-1-phenyl-2-butenone; R = H

Fig. 6 Structures **87–96**

berkedrimanes A and B (**85–86**) (Fig. 6) were isolated by Steirle and coworkers from *P. solitum*, a fungus obtained from the secreted slime of acidophilic yeast found at the surface of the Pit lake (Stierle et al. 2012b). Berkedrimane B **86** is the hydroxy derivative of berkedrimane A **85**, which exhibited inhibitory activity against casp-1 and another important protease involved in apoptosis, caspase-3 (casp-3), with IC_{50} values of 255 and 511 μM, respectively. Berkedrimane B **86** was a slightly more potent inhibitor of casp-1 and casp-3 with IC_{50} values of 221 and 123 μM, respectively. In addition, compounds **85–86** both moderately inhibited casp-3 with IC_{25} values of 15 and 10 μM, respectively. Furthermore, both compounds inhibited the production of IL-1β in titanium nanowire- and eubacterial LPS-induced THP-1 cells at micromolar concentrations, demonstrating that these compounds may be useful in the development of anti-inflammatory agents that target the inflammasome, the multiprotein complex involved in providing innate immunity via the activation of proinflammatory cytokines IL-1β and IL-18.

The Berkeley Pit continues to be a unique and dynamic ecosystem inhabited by acidophiles with interesting metabolic profiles that have not been fully explored. This pit could be a rich source of the most lucrative pharmacophores to ever to be mined, as more than 20 metabolites have been reportedly produced from microorganisms isolated from this niche. *Penicillium* species from other acidic environments have also been reported to produce novel bioactive agents. For example, Wang and coworkers isolated an acid-tolerant strain of *P. purpurogenum* JS03-21 from the local red soil used for manufacturing purple pottery in Jianshui, Yunnan, China (Wang et al. 2011b). When cultured at pH 2, *P. purpurogenum* produced six new compounds, purpurquinones A–C (**87–89**), purpuresters A and B (**90–91**), and the napthoquinone 2,6,7-trihydroxy-3-methylnapthalene-1,4-dione **92** (Fig. 6). Purpurquinones A–C (**87–89**) share an azaphilone core, but have different patterns of hydroxylation. Purpuresters A and B (90–91) are isobenzofuranones with different substitutions at C3 on the pentasubstituted aryl ring. All compounds were tested for anti-viral activity against influenza A virus (H1N1) and purpurquinones B and C (88–89) as well as purpurester A 90 exhibited stronger anti-influenza virus activity compared to that of the ribavirin positive control (IC$_{50}$, 100.8 µM) with IC$_{50}$ values of 61.3, 64.0, and 85.3 µM, respectively. This study demonstrated that simply changing the pH of fermentation media is a promising technique that can be used for drug discovery, as the anti-viral purpurester A 90 and purpurquinones B and C (88–89) were only produced at acidic pH.

Another acidophilic eubacterial strain, *Lysinibacillus fusiformis* KMCOO3, was recently isolated by Park and coworkers from acidic (pH 3) coal mine drainage contaminated with sulfuric acid and iron-rich, heavy-metal ions (Park et al. 2012). The authors screened the culture broth of this eubacterium to identify agents that may be used as part of its unique, offensive, and defensive mechanisms when under ecological pressure. *L. fusiformis* KMCOO3 was determined to produce the unusual spiro-cyclopentenones spirobacillenes A and B (93–94) as well as (Z)-3-hydroxy-4-(3-indoyl)-1-hydroxyphenyl-2-butenone 95 and (Z)-3-hydroxy-4-(3-indoyl)-1-phenyl-2-butenone 96 (Fig. 6). Spirobacillenes A and B (93–94) both possess spiro-cyclopentenones and unique indole and indolenine moieties, respectively. The C–C bond between the spiro[4.5]decane moiety in spirobacillene A 93 and the highly functionalized spiro-cyclopentenone in spirobacillene B 94 is novel feature for naturally occurring molecules. With the exception of (Z)-3-hydroxy-4-(3-indoyl)-1-phenyl-2-butenone 96, these molecules were tested for antimicrobial, cytotoxic, and anti-inflammatory activity. Only (Z)-3-hydroxy-4-(3-indoyl)-1-hydroxyphenyl-2-butenone 95 exhibited antimicrobial activity against *M. luteus*, *Enterococcus hirae*, and *S. aureus* with MIC values of 3.13 µg/ml, 3.1.3 µg/ml, and 12.5 µg/ml, respectively. In addition, spirobacillene A 93 weakly inhibited reactive oxygen species and NO production with IC$_{50}$ values of 39 µM and 43 µM, respectively. *L. fusiformis* is a prime example of how extensive modifications of solely the indole scaffold by eubacterial biosynthetic machinery can result in novel, bioactive spirocyclic molecules.

The Berkeley Pit represents one of the many acidic niches found on Earth, and the prolific number of bioactive metabolites isolated from a single *P. rubrum* isolate demonstrates the unlimited potential for finding new pharmacophores in these locations. Several screening programs have been initiated to identify new secondary metabolite producers from acidic environments. For example, because the production of secondary metabolites is not a uniform trait across all organisms that belong to the order of *Actinomycetales*, Busti and coworkers have started to study uncultured actinomycetes that share several characteristics as those of secondary metabolite-producing *Streptomyces* to identify new prolific producers of antibiotics (Busti et al. 2006). The authors isolated several filamentous actinomycetes from mildly acidic soils obtained from temperate forests in Gerenzo, Italy, (pH 4.3) and Nicaragua (pH 5.0). Genomic sequencing of these strains revealed the presence of chromosomes that are 8 Mb or larger as well as their potential to produce natural products, as several genes encoding type I and type II polyketide

Table 3 Bioactive compounds isolated from terrestrial acidophiles

Acidophile	Optimal pH	Source	Compound	Bioactivity	References
Penicillium rubrum	2.7	Lake sediment	Berkeleydione **67**	MMP-3 and casp-1 inhibitor; cytotoxic	Stierle et al. (2004)
P. rubrum	2.7	Lake sediment	Berkeleytrione **68**	MMP-3 and casp-1 inhibitor	Stierle et al. (2004)
P. rubrum	2.7	Lake sediment	Berkeleyones A–C (**69–71**)	Casp-1 inhibitor	Stierle et al. (2011)
Chlorella mutabilis-associated *Penicillium* sp.	2.7	Lake sediment	Berkelic acid **72**	MMP-3 and casp-1 inhibitor; cytotoxic	Stierle et al. (2006)
P. rubrum; *Penicillium* sp. IMU-0035	2.7	Lake sediment; soil	Berkeleyacetals A–C (**73–75**)	Interleukin-1 receptor-associated kinase-4; MMP and casp-1 inhibitor (**75**); cytotoxic (**75**)	Stierle et al. (2007), Etoh et al. (2013)
P. rubrum	2.7	Lake sediment	Berkeleyamides A–D (**76–79**)	MMP and casp-1 inhibitor	Stierle et al. (2008)
P. rubrum	2.7	Lake sediment	Berkazaphilones A–C (**80–81; 84**)	Casp-1 inhibitor	Stierle et al. (2012)
P. rubrum	2.7	Lake sediment	Berkedienoic acid **82**	N/A	Stierle et al. (2012)
P. rubrum	2.7	Lake sediment	Berkedienolactone **83**	N/A	Stierle et al. (2012)
P. solitum	2.7	Lake surface	Berkedrimanes A and B (**85–86**)	Casp-1 and casp-3 inhibitor	Dimise et al. (2008)
P. purpurogenum	2	Soil	Purpurquinones A–C (**87–89**)	Anti-viral against influenza A (**88–89**)	Wang et al. (2011b)
P. purpurogenum	2	Soil	Purpuresters A and B (**90–91**)	Anti-viral against influenza A (**90**)	Wang et al. (2011b)
P. purpurogenum	2	Soil	2,6,7-Trihydroxy-3-methylnapthalene-1,4-dione **92**	N/A	Wang et al. (2011b)

(continued)

Table 3 (continued)

Acidophile	Optimal pH	Source	Compound	Bioactivity	References
Lysinibacillus fusiformis KMCOO3	7 (pH 3)*	Soil	Spirobacillenes A and B (**93–94**)	Weak inhibitor of reactive oxygen species and NO production (**93**)	Park et al. (2012)
L. fusiformis KMCOO3	7 (pH 3)*	Soil	(*Z*)-3-hydroxy-4-(3-indoyl)-1-hydroxyphenyl-2-butenone **95**	Antimicrobial	Park et al. (2012)
L. fusiformis KMCOO3	7 (pH 3)*	Soil	(*Z*)-3-hydroxy-4-(3-indoyl)-1-phenyl-2-butenone **96**	N/A	Park et al. (2012)

N/A information not available

* pH of the environment from which the microbe was isolated

synthases and nonribosomal peptide synthetases were amplified by PCR. In addition, the authors showed the antibiotic-producing potential of these slow-growing acidophiles that belong to the genera *Catenulispora* and *Actinospica*, new genera of previously uncultured actinomycetes, which share several features with *Streptomycetes*. Not only do these acid-tolerant microorganisms have the potential to produce new metabolites, but they also have the potential to known antibiotics reported from other unrelated eubacteria, such as isochromanequinone. The authors proposed that in order to find new antibiotics, strains belonging to new taxa should fit the following criteria: (1) have the ability to produce secondary metabolites; (2) have diversified secondary biosynthetic pathways; (3) be genetically diverse; (4) be able to obtained in large numbers; and (5) be amenable for large-scale fermentation. Criteria 1, 2, possibly 3, and 4 were demonstrated in this study, and the authors plan to expand this to other microorganisms that belong to novel actinomycete taxa.

A summary of the bioactive agents, optimal pH, and microbial sources described in this chapter is shown in Table 3. This list will increase over time as the number of screens for finding optimal bioleaching acidophiles increases due to the growing needs of industry. Furthermore, over the past 15–20 years, more patents have been filed by the Diversa Corporation to study acidophiles for enzymes and bioactivity (Short 2000; Short and Keller 2005). In the United States alone, there are still several EPA Superfund sites harboring acidophiles that remain to be explored, including the Gowanus Canal in New York, Portland Harbor in Oregon, Taylor Lumber and Treating in Oregon, Hudson River in New York, Pearl Harbor Naval Complex in Hawaii, and Tar Creek in Oklahoma. With the molecular tools to determine the taxa of microorganisms as well as their biosynthetic genes, researchers will be able to identify new microorganisms with the potential to produce novel new pharmacophores. In addition, modifying the pH of culture conditions may produce many unusual chemical entities.

5 Alkaliphiles

In Sect. 4, we described how the pH of an environment could significantly influence the growth and metabolome of a microorganism. Not only are there microorganisms that thrive at acidic pH, but some grow well in alkaline environments (pH > 10) called alkaliphiles. Microorganisms that grow and reproduce at alkaline pH but grow optimally at neutral pH are categorized as being alkali tolerant. In 1956, alkali-tolerant eubacteria were first reported by Koki Horikoshi, a graduate student in the Department of Agricultural Chemistry at the University of Tokyo, who noticed that the *Aspergillus oryzae* that he was growing was no longer present in the culture flask, which smelled of ammonia and contained pH 9 media. Horikoshi later determined that a lytic microorganism, *Bacillus circulans*, was growing in his culture flask. Up to this point in time, little microbiological research had been conducted at alkaline pH because alkaline foods were uncommon. However, after incubating eubacteria isolated from soil collected from various locations around his school in alkaline fermentation media at 37 °C, several microbes grew. Horikoshi then spent the next four decades studying how these organisms survived at basic pH (mostly at pH 10) and maintain their intracellular environments at pH 7–8 (Horikoshi and Bull 2011).

Alkaliphiles are considered to be unique extremophiles because they can exist in traditional agricultural environments (e.g., alkaline soil and manure) as well as alkaline environments, such as soda lakes, underground alkaline water, man-made alkaline environments, and alkaline niches within other organisms. Like acidophiles, alkaliphiles have developed mechanisms to maintain a pH gradient across the cell membrane. Alkaliphiles have a lower intracellular pH by maintaining an electrochemical gradient of protons across the cell membrane and having diverse ion transporters, channels, and enzymes on the cell surface to help accumulate cytoplasmic protons. If the cytoplasmic pH of nonalkaliphilic *E. coli* increases to ≥ 8, then the cells die. Interestingly, the cytoplasm of alkaliphiles can tolerate growing at a higher pH, indicating that there are cytoplasmic processes, such as cell division, and enzymes, designed to function optimally under these conditions.

Alkaliphiles have mainly had a significant impact on industrial applications due to their enzymes having various unique properties, such as thermostability, barophilicity, and psychrophilicity. In 2011, these enzymes had a dominant position in the approximately \$5.1 billion enzyme industry worldwide as stable constituents of detergents as well as additives used in animal feed processing, bioremediation, food supplements, the finishing of fabric, and in the pulp and paper industries (Sarethy et al. 2011). While the popularity of alkaliphiles stems from their unique enzymatic properties, a few have been reported to produce bioactive secondary metabolites. The small number of alkaliphiles that produce new small molecules may be related to the fact that they have not been extensively explored due to the instability of compounds at alkaline pH and the lower amount of alkaline food.

Some of the first secondary metabolites to be isolated from an alkaliphile were the peptide antibiotics 1907-II **97** and 1907-VIII **98** (Fig. 7) (Sato et al. 1980). Antibiotic 1907-II **97** is a peptide consisting of β-hydroxyleucine, 4-methylproline, three α-aminoisobutyric acids, two leucines, β-alanine, 2-amino-4-methyl-6-hydroxy-8-oxodecanoic acid, and 4-methyl-2-hexenoic acid, whereas 1907-VIII **98** is a methylated analog of 1907-II **97**. These antibiotics were only produced by *Paecilomyces lilacinus* 1907, a microorganism that grows in alkaline media with a pH of 9–10.5. Notably, 1907-II **97** and 1907-VIII **98** exhibited antimicrobial activity against eubacteria and fungi with MIC values ranging from 0.78–50 to 0.78–50 µg/ml, respectively.

While screening the supernatants of alkaliphilic soil eubacterial cultures for novel inhibitors of aldose reductase (EC 1, 1, 1, 21), an enzyme implicated in the development of various diabetic complications, Bahn et al. isolated *Corynebacterium* sp. YUA25 from the culture broth of Korean soil (Bahn et al. 1998). YUA001 **99** is an *N*-substituted tyramine that exhibited weak inhibitory activity against aldose reductase with an IC_{50} value of 1.8 mM compared to that of the tolrestat positive control (IC_{50}, 16 µM). The addition of an alkenyl

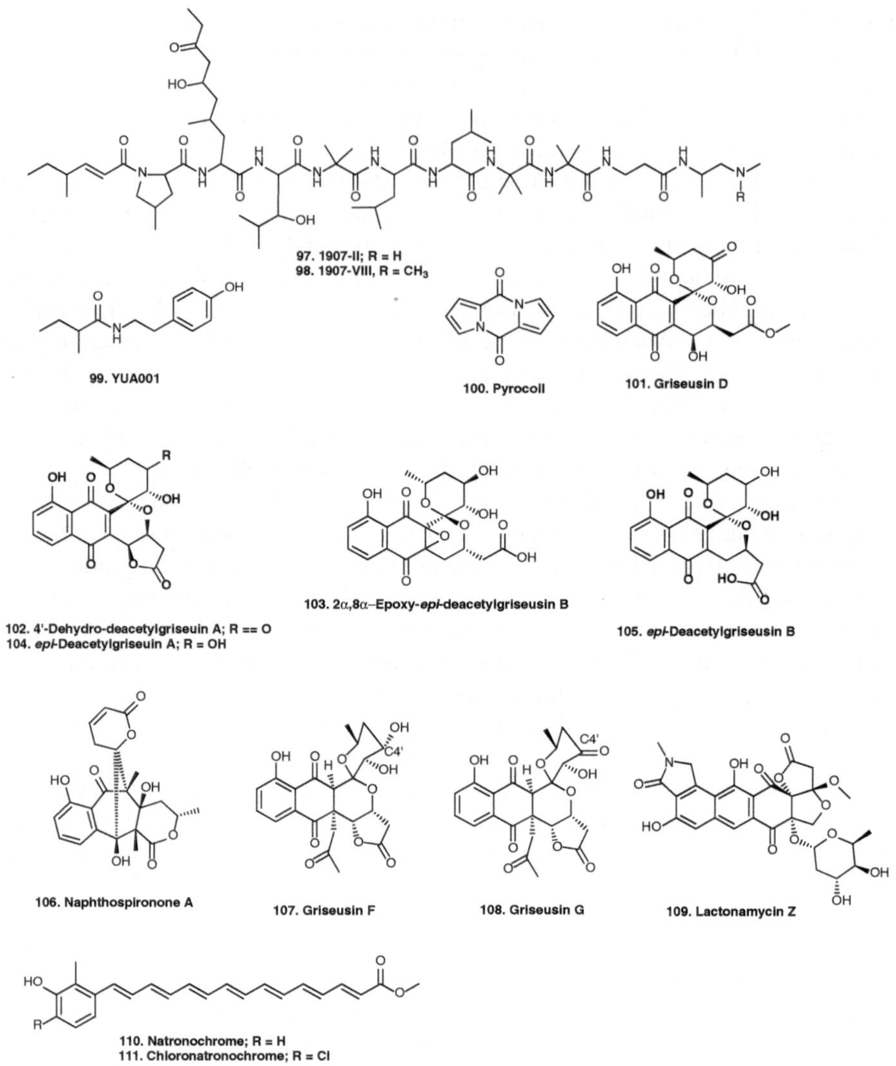

Fig. 7 Structures **97–111**

substituent to YUA001 **99** appeared to have an effect on inhibition, as *N*-2-*p*-hydroxyphenylethyl maleamic acid exhibited 22-fold more potent inhibitory activity against aldose reductase with an IC$_{50}$ value of 80 µM, which is the same order of magnitude as that of the positive control (Sun et al. 2001). Thus, the isolation of new pharmacophores from natural sources could possibly serve as leads to generate synthetic analogs with improved bioactivity.

During a screening program to identify novel secondary metabolites from alkaliphilic and alkali-tolerant actinomycetes, Dietera and coworkers isolated the diketopiperazine pyrocoll **100** (Fig. 7), a synthetic constituent of cigarette smoke that had previously never been isolated from a natural source (Dietera et al. 2003). Pyrocoll **100** was isolated from the alkaliphilic *Streptomyces* sp. AK409 obtained from steel waste tip soil. Interestingly, pyrocoll **100** exhibited antimicrobial activity against *Arthrobacter* strains (MIC, 1–10 μg/ml) and yeast, such as *Rhodococcus erythropolis* DSM 1069 (MIC, 10 μg/ml). Weak antifungal activity was observed against filamentous fungi, such as *Botrytis cinera*, *A. viridi nutans*, and *Paecilomyces variotii*. Pyrocoll **100** also inhibited the growth of gastric adenocarcinoma HMO2, hepatocellular carcinoma HepG2, and breast carcinoma MCF-7 cells with ED_{50} values of 1.5, 2.3, and 12 μM, respectively. Furthermore, compound **100** selectively inhibited the growth of mammalian L5178y lymphoma cells and HeLa S3 cells with ED_{50} values of 3.5 and 11 μM, respectively. Pyrocoll **100** also displayed moderate antimalarial activity (IC_{50}, 6.4 μM) against a pathogenic agent of malaria, *Plasmodium falciparum*, compared to that of the chloroquine positive control (IC_{50}, 0.24 μM). Compound **100** has been reported to have other bioactivity, such as weak antiparasitic activity against the protozoan parasites *Trypanosoma cruzi* (IC_{50}, 95 μM) and *T. brucei rhodesiense* (IC_{50}, 11 μM) compared to that of the positive controls benznidazole (IC_{50}, 4.9 μM) and melarsoprol (IC_{50}, 0.003 μM), respectively. Synthetic pyrocoll **100** has been used as a probe to understand the effects of diabetes on calcium channels, such as type-2 ryanodine receptor calcium-release channels (RyR2), a protein posttranslationally affected by diabetes before its expression is decreased altogether (Bidasee et al. 2003). These data demonstrate how we can find new sources of naturally occurring molecules that were previously only accessible via synthesis and use these molecules to study the progression of disease.

Bacterial strains belonging to the *Nocardiopsis* genus are ecologically diverse and ubiquitously distributed in various environments, including saline and alkaline environments. In a screen to find novel cytotoxic agents, a new pyranonaphthoquinone, griseusin D **101** (Fig. 7), was isolated by Li and coworkers from the alkaliphilic *Nocardiopsis* sp. YIM 80133 (Li et al. 2007). This eubacterium was isolated from soil collected from Qinghai Province, China, and grown in pH 9.5 production media. Griseusin D **101** has a naphthoquinone core with a fused spiroacetal ring system (Tsuji et al. 1976). Compound **101** exhibited cytotoxic activity against human leukemia and lung adenocarcinoma cells with IC_{50} values of 0.52 and 44 μM, respectively. Interestingly, other cytotoxic griseusin derivatives were later identified from alkaliphilic *Nocardiopsis* species obtained from polluted Chinese soil.

Similar to the work done by Stierle et al. on bioprospecting of the Berkeley Pit, He and coworkers screened microorganisms isolated from the alkaline soil of the Datun tin mine tailings in Yunnan province, China, for new metabolites (He et al. 2007). Six tumor cell lines, large cell lung cancer LXFL 529L, mammary cancer MAXF 401NL, melanoma MEXF 462NL, uterine cancer UXF 1138L, gastric cancer GXF251L, and renal cancer RXF 486L cell lines, with various chemosensitivity toward commonly used chemotherapeutic agents were used to screen microbial extracts for bioactivity. Of the active extracts, alkaliphilic

Nocardiopsis sp. YIM80133 and DSM1664 strains were identified as the producers of new pyranonaphthoquinone griseusins, 4′-dehydro-deacetylgriseusin A **102**, 2α,8α-epoxy-*epi*-deacetylgriseusin B **103**, *epi*-deacetylgriseusin A **104**, and *epi*-deacetylgriseusin B **105** (Fig. 7). These griseusins possess core pyranonaphthoquinone structures with a polyketide fragment decorated with either a fused lactone ring, hydroxyl groups, or an epoxy moiety. With the exception of 2α,8α-epoxy-*epi*-deacetylgriseusin B **103** and *epi*-deacetylgriseusin B **105**, all isolated compounds inhibited the growth of the six tumor cell lines. The average IC_{50} values for 4′-dehydro-deacetylgriseusin A **102** and *epi*-deacetylgriseusin A **104** against the six cell lines were 0.392 and 5.32 μM, respectively. Further analysis of the most cytotoxic agent, compound **102**, against monolayer cultures of 37 different human tumor cell lines revealed that 4′-dehydro-deacetylgriseusin A **102** exhibited potent cytotoxic activity with an average IC_{50} value of 430 nM. 4′-Dehydro-deacetylgriseusin A **102** was selective against various breast cancer (MDA-MS 231, MDA-MB 468, MCF-7, MAX7 401; IC_{50}, 150–345 nM), renal cancer (RXF 393NL, RXF 468L, RXF 944L; IC_{50}, 95–250 nM), and melanoma (MEXF 276L, MEXF 394NL, MEXF 462NL, MEXF 514L, MEXF 520L; IC_{50}, 87–280 nM) cell lines. In addition, in clonogenic assays with established human tumor xenografts used to identify candidate tumors for in vivo studies, 4′-dehydro-deacetylgriseusin A **102** selectively inhibited the colony formation of a variety of tumors, including colon cancer, breast cancer, melanoma, pancreatic cancer, renal cancer, and leukemia solid tumors in semisolid medium. Based on the mode of action of other known antitumor therapeutics, the authors speculate that this cytotoxic agent may have an unusual mechanism of action.

Nocardiopsis sp. strain YIM DT266 was also isolated from the same site and reported to produce the unique cytotoxic agent naphthospironone A **106** (Fig. 7) (Ding et al. 2010). This molecule contains an unusual spiro[bicycle[3.2.1]octane-pyran]dione ring system and exhibited very weak cytotoxic activity against murine fibrosarcoma L929, HeLa, and human lung cancer AGZY cells with IC_{50} values within the range of 80–221 μM. Naphthospirone A **106** also exhibited moderate to weak antimicrobial activity against *B. subtilis*, *S. aureus*, *E. coli*, and *A. niger* with MIC values within the range of 11–25 μg/ml. Two unprecedented 2α-hydro-8α-(2-oxopropyl)-substituted spiro-naphthoquinones with an unusual C23 polyketide skeleton, griseusins F and G (**107–108**) (Fig. 7), were also isolated from the same eubacterium (Ding et al. 2012). Griseusin G **108** is an oxidized derivative of griseusin F **107** by possessing a ketone moiety instead of a hydroxyl group at the C4′ position. Griseusins F and G (**107–108**) both exhibited potent cytotoxicity against human melanoma B16, breast carcinoma MDA-MB-435S, pancreatic cancer CFPAC-1, renal carcinoma ACHN, and colorectal carcinoma HCT 116 cell lines with IC_{50} values ranging from 0.37 to 0.82 μM, which are on the same order of magnitude as those of the cisplatin positive control. Both compounds also exhibited potent antibiotic activity against *S. aureus* ATCC 29213, *M. luteus*, and *B. subtilis* with MIC values ranging from 0.80 to 1.65 μg/ml. This wide range of bioactivity is not surprising because griseusins have been reported to exhibit antibiotic activity against methicillin-resistant *S. aureus* as well as cytotoxic activity

(Tsuji et al. 1976; Li et al. 2007; Igarashi et al. 1995). These studies further demonstrate how valuable pharmaceutical leads can be found in the most polluted soils.

Several other screening programs have been initiated to search for new antibiotics produced by alkaliphiles. The antibiotic and antitumor agent lactonamycin Z **109** (Fig. 7) was isolated by Höltzel and coworkers from the alkaliphilic *S. sanglieri* AK 409 (Höltzel et al. 2003). This alkaliphile was isolated from pine wood soil obtained from Hamsterley Forest, County Durham, UK. Lactonamycin Z **109** has unique structural features that include both a naphtha[*e*]isoindole ring system and an oxygen-rich perhydrofuran–furanone ring system with a labile tertiary methoxy group. Compound **109** exhibited weak antibiotic activity against Gram-positive eubacteria with inhibition zones from 7 to 24 mm at a concentration of 1 mg/ml. In addition, lactonamycin Z **109** inhibited the growth of gastric adenocarcinoma HMO2, breast carcinoma MCF-7, and hepatocellular carcinoma HepG2 cell lines in the G2/M cell cycle phase with GI_{50} values of 3.2, 1.5, and 8.7 μM, respectively. This bioactivity is not surprising as lactonamycin, which has a modified sugar moiety, exhibited cytotoxic activity against a variety of leukemia, sarcoma, carcinoma, fibrosarcoma, and melanoma cell lines with IC_{50} values within the range of 0.11–5.79 μM (Matsumoto et al. 1999). Lactonamycin also exhibited potent antibiotic activity against various multidrug-resistant eubacteria, indicating that the modified sugar may be important for antibacterial activity. The biosynthetic gene clusters for lactonamycin and lactonamycin Z **109** have been identified, revealing unusual polyketide starter units of glycine or a glycine derivative extended by nine acetate units, which could be exploited to engineer the production of new chemical derivatives (Zhang et al. 2008).

As more sequenced genomes become available, more novel antibiotics and other bioactive metabolites will be isolated from alkaliphiles. For example, Lawton and coworkers sequenced the entire genome of the alkaliphilic eubacterium *B. halodurans* C-125 isolated from soil and identified genes involved in the biosynthesis of a lantibiotic (Lawton et al. 2007). Using assay-guided fractionation, haloduracin, a complex two-peptide lantibiotic composed of posttranslationally modified Halα and Halβ, was observed to have antimicrobial activity against *Lactococcus lactis* HP and inhibit the spore outgrowth of *B. anthacis*. Antibiotic activity was further confirmed by mutagenesis of the Hal genes involved in the expression of these peptides. This was the first report of the production of a two-peptide lantibiotic produced from an alkaliphilic *Bacillus* species. Furthermore, the mode of action of haloduracin was determined and compared to that of the prototypical two-peptide lantibiotic, lacticin 3147, whereby the α-peptide binds to a target on the cell surface of Gram-positive eubacteria and the β-peptide permeabilizes the membrane, inducing an efflux of potassium (Oman and van der Donk 2009). With the inexpensive costs of genome sequencing, we foresee an increase in the number of publications on the identification of new bioactive, small molecules produced from mining the genome of alkaliphiles.

Modifications to cell culture conditions are standard microbiological protocols for activating the biosynthesis of new molecules and can be applied to alkaliphiles and other extremophiles. For example, the temperature of culture conditions was reported to affect the metabolic profile of phenazine antibiotics produced by an alkaliphilic actinomycete, *N. dassonvillei* strain OPC-15, isolated from a soil sample collected in Mino City Osaka, Japan (Tsujibo et al. 1988). Under alkaline conditions (pH 10), 1,6-dihydroxyphenazine accumulated and slowly converted to 1,6-dihydroxyphenazine 5-monooxide when grown at 27 °C. However, when the temperatures were shifted from 27 to 4 °C, 1,6-dihydroxyphenazine 5-monooxide converted to 1,6-dihydroxyphenazine 5,10-dioxide.

Applying Na^+ stress on a microorganism may also be useful strategy to trigger the biosynthesis of new bioactive metabolites that are either not produced or produced in trace amounts. Pomati and coworkers varied the salt concentrations of the alkaliphilic cyanobacterium *Cylindrospermopsis raciborskii* strain T3 and observed an increase in the potent shellfish toxin saxitoxin (Pomati et al. 2004). Saxitoxin selectively blocks voltage-gated Na^+ channels, which causes death in animals. Increasing the concentration of salt increases the influx of Na^+ into the cell, which has a significant effect on the Na^+/H^+ antiporters that maintain the pH gradient across the cell membrane, as protons are exchanged for cytoplasmic Na^+. In this study, increasing the salt concentration in media containing *C. raciborskii* triggered the production of saxitoxin to maintain cell homeostasis (i.e., reduce intracellular Na^+). The effects of Na^+ stress on alkaliphiles have also been demonstrated by Takaichi et al. on the isolation of novel pigments with antioxidant properties (Takaichi et al. 2004). The authors extracted two new pigments, natronochrome **110** and chloronatronochrome **111**, produced by the alkaliphilic γ-proteobacterium *Thialkalivibrio versutus* strain ALJ isolated from the hypersaline soda lakes in Kenya. Natronochrome **110** is a 23-carbon polyene chain with terminal methyl ester and a methylated and hydroxylated phenol moiety, and chloronatronochrome **111** is a chlorinated analog of natronochrome **110** (Takaichi et al. 2004). Interestingly, these pigments were not produced under low salt conditions (Banciu et al. 2005). No bioactivity has been reported for either of these compounds.

Although Table 4 summarizes the bioactive agents discussed in this chapter as well as their sources and optimal pH required for growth, this list will significantly grow in time, as increased screening efforts over the past decade have demonstrated that researchers have underestimated the potential of alkaliphiles to produce pharmaceutical leads. An excellent example of this is the recent study by Otto and coworkers in which various eubacterial strains, including extremophiles and mesophiles, were screened for their ability to hydrolyze organophosphorus nerve agents [VX, VR, and soman (GD)] with the goal of developing catalytic bioscavengers to serve as prophylactics (Otto et al. 2013). Microorganisms, including two strains of alkaliphiles, with proteins bearing any similarity (46–51 %) to the protein sequence of organophosporus hydrolase from *Brevundimonas diminuta* were screened for activity, and the alkaliphile *Ammoniphilus oxalaticus* was identified to be one of the three strains with significant activity. Once the biologic agent(s) is isolated, it could serve as a lead for the design of catalytic bioscavengers of G- and V-agents administered prior to nerve agent exposure. Other biological screens, such as the work by Thumar et al. and Zitouni et al. on alkaliphiles and alkali-tolerant actinomycetes from desert soil (Thumar et al. 2010; Zitouni et al. 2005), and the (meta)

Table 4 Bioactive compounds isolated from terrestrial alkaliphiles

Alkaliphile	Optimal pH	Source	Compound	Bioactivity	References
Paecilomyces lilacinus 1907	9–10.5	N/A	1907-II **97**	Antimicrobial	Sato et al. (1980)
P. lilacinus 1907	9–10.5	N/A	1907-VIII **98**	Antimicrobial	Sato et al. (1980)
Corynebacterium sp. YUA25	N/A	Soil	YUA001 **99**	Very weak aldose reductase inhibitor	Bahn et al. (1998)
Streptomyces sp. AK409	9	Soil	Pyrocoll **100**	Antimicrobial; cytotoxic; antimalarial; weak antiparasitic; (de)activator of type 2 receptor calcium-release channels	Dietera et al. (2003), Bidasee et al. (2003)
Nocardiopsis sp. YIM 80133; *Nocardiopsis* sp. DSM1664	9.5–10.5	Soil	Griseusin D **101**	Cytotoxic	Li et al. (2007)
Nocardiopsis sp. YIM 80133; *Nocardiopsis* sp. DSM1664	9.5–10.5	Soil	4'-Dehydro-deacetylgriseusin A **102**	Cytotoxic	He et al. (2007)
Nocardiopsis sp. YIM 80133; *Nocardiopsis* sp. DSM1664	9.5–10.5	Soil	2α,8α-Epoxy-*epi*-deacetylgriseusin B **103**	N/A	He et al. (2007)
Nocardiopsis sp. YIM 80133; *Nocardiopsis* sp. DSM1664	9.5–10.5	Soil	*epi*-Deacetylgriseusin A **104**	Cytotoxic	He et al. (2007)
Nocardiopsis sp. YIM 80133; *Nocardiopsis* sp. DSM1664	9.5–10.5	Soil	*epi*-Deacetylgriseusin B **105**	N/A	He et al. (2007)
Nocardiopsis sp. strain YIM DT266	12; * (pH 10)	Soil	Naphthospironone A **106**	Weakly cytotoxic; antimicrobial	Ding et al. (2010)
Nocardiopsis sp. YIM DT266	12; * (pH 10)	Soil	Griseusins F and G (**107–108**)	Antimicrobial; cytotoxic	Ding et al. (2012)
Streptomyces sanglieri AK 409	7; ** (pH 9)	Soil	Lactonamycin Z **109**	Antimicrobial; cytotoxic	Höltzel et al. (2003), Matsumoto et al. (1999)

(continued)

Table 4 (continued)

Alkaliphile	Optimal pH	Source	Compound	Bioactivity	References
Thialkalivibrio versutus strain ALJ 15	10	Soda lake	Natronochrome **110**	N/A	Takaichi et al. (2004)
T. versutus strain ALJ 15	10	Soda lake	Chloronatronochrome **111**	N/A	Takaichi et al. (2004)

N/A information not available

* pH of the environment the microbe was isolated from

** Optimal pH for microbial growth

genomic sequencing (Foti et al. 2008; Siddaramappa et al. 2012; Takami et al. 2000; Keshri et al. 2013) of alkaliphiles from different environments have revealed diverse and novel taxon and genes, many of which are unexplored sources of new chemical entities.

6 Halophiles

Extreme halophiles are microorganisms that grow in environments with 15 % (w/v) salt to saturation, which are concentrations considered to be hostile to most organisms. Moderate halophiles grow optimally in saline environments with salt concentrations ranging from 3 % (w/v) (seawater) to 15 % (w/v) (Mancinelli 2005). Microorganisms that prefer nonsaline environments but grow in environments containing up to 10 % salt are considered to be halotolerant. Halophilic eubacteria inhabit hypersaline or thassohaline environments, such as salt lakes (e.g., Great Salt Lake, Utah, USA, and the Dead Sea), salterns (artificial salt lakes), saline soils, man-made environments (e.g., the division of a lake by a rock-fill railroad causeway in the 1950s), the ocean and sea, underground deposits of rock salt, and salted food products. As 70 % of the planet's surface is covered with salty seawater, many marine microorganisms, by our definition, are classified as halophiles and will be discussed in the second book in this series.

Halophiles are highly diverse aerobic and anaerobic microorganisms, and those that survive in sodium-saturated brines (>300 g/ml) are dispersed among the three-domain *Archaea–Eubacteria–Eukarya* phylogenetic tree based on small rRNA comparisons. The microbial flora of thassohaline environments has been studied for years, but only recently have scientists been able to culture the most dominant species. The most widespread representatives of *Archaea, Eubacteria,* and *Eukarya* in thassohaline environments are microorganisms belonging to the orders *Halobacteriales* and *Haloquadratum; Salinibacter* and *Halanaerobiales*; and the genus *Dunaliella*, respectively. In Sect. 5, increased salt concentrations in some alkaliphile cultures were considered to be deleterious. On the other hand, lower salt concentrations promote cell lysis. To counterbalance external osmotic pressure, halophiles maintain their turgor by accumulating inorganic ions, such as potassium chloride, in the cytosol such that their internal salt concentration is always comparable to that of their external environment (Pflüger and Müller 2004). Osmolytes or organic molecules are also produced to increase turgor pressure, while salt is eliminated from the cytosol. Sodium ions are excluded from the cytoplasm as much as possible via Na^+/H^+ antiporter systems. As a result, proteins have adapted to function optimally at high salt concentrations. For example, many enzymatic reactions are resistant to Na^+ and require high concentrations of K^+ to increase protein stability and adopt the appropriate conformation required for enzymatic catalysis (Vellieux et al. 2007). Negatively charged amino acid residues also cover the surface of halophilic proteins to increase their solubility in saturated sodium environments.

As the salt concentration of an environment increases, some metabolic activities do not take place, such as autotrophic nitrification, which is involved in the oxidation steps in the conversion of ammonium to nitrate, and the number of microorganisms in that environment decreases. However, other processes still occur, such as denitrification, anoxygenic photosynthesis using sulfide as an electron donor, and aerobic respiration. While metabolic processes are limited under hypersaline conditions, halophiles are rich sources of enzymes that function in high salt environments, poly-β-hydroxyalkanoate, exopolysaccharides, β-carotene (strains of *Duniella*), retinal proteins (e.g., rhodopsins) for biosensor applications, and glycerol and ectoine, which are used as stabilizing agents for sensitive enzymes (Oren 2010). Although halophiles have several useful applications, some of which have been used for centuries, they have not been as exploited as other extremophiles used in the production of solar salt, fermented foods, and preservative agents in the food and leather industries.

Several novel bioactive proteins and peptides have been isolated and (partially) characterized from halophiles, including the halocins and microcins. In 1982, one *Halococcus* strain and 39 strains of halophilic rods of the domain *Archaea* (haloarchaea) were screened against each other, and seven strains were determined to have antagonistic activity and produce proteinaceous antibiotics called halocins. These proteins/peptides have molecular weights ranging from 2.5 to 35 kDa and are universally produced by archaea that typically inhabit environments containing 4–5 M salt (Karthikeyan et al. 2013). The structures of these peptide/proteins are not provided, as they have not all been fully defined.

Halocins exhibit a narrow range of activity only inhibiting the growth of several haloarchaea and not the producing microorganism (Rodriguez-Valera et al. 1982). However, the antimicrobial activity of halocins is broad with respect to inhibiting the growth of other haloarchaea. Notably, these protein/peptide sequences do not resemble any sequence in protein sequence databases. When resources are limited, halocins lyse competitors to supplement the nutrients of their producing organism. In fact, most thassohaline environments are inhabited by haloarchaea due to the stability and function of halocins under saturated saline conditions, which may explain why there is so little species diversity among these archaea. The functions of these antibiotics are similar to those of bacteriocins produced by eubacteria to control eubacterial populations and may evolutionarily related (Riley 1998).

The following studies on haloarchaea have demonstrated the bioactivity of halocins: (1) Meseguer et al. screened 79 haloarchaeal strains for antagonist interactions between strains and observed growth inhibition, with the exception of two strains, due to the production of halocins (Meseguer et al. 1986); (2) Torreblanca and coworkers also observed a wide range of antagonistic activity among 68 haloarchaeal strains, with the exception of only one strain, due to the production of halocins (Torreblanca et al. 1994); and (3) Platas and coworkers purified a 31-kDa halocin H1 from the archaeal strain *Haloferax mediterranei* M2a and observed its inhibition of other strains of *Halobacterium salinarum* at salt concentrations higher than 1.5 M (Platas et al. 2002).

Since 1982, the following halocins or smaller microcins have been identified from various haloarchaea (O'Connor and Shand 2002): halocins H1, H2, H3, H4

(28 kDa) (Meseguer and Rodriguez-Valera 1985), H5, H7 (formerly known as H6) (Torreblanca et al. 1989), S8 (2.5-kDa; microcin) (Price and Shand 2000), C8 (6.3-kDa) (Li et al. 2003), R2 (6.2-kDa; microcin) (Rdest and Sturm 1987), R2 (6.2-kDa; microcin) (Rdest and Sturm 1987), G1, A4, Sech7a (Pašic et al. 2008), and SH10 (Platas et al. 2002; Rodriguez-Valera et al. 1982). Only a few of these halocins have been isolated, characterized, and determined to be salt dependent (Shand and Leyva 2007). The mechanism of action of halocins/microcins remains to be fully elucidated but appears to be related to disturbing cell permeability of the flux of ions across a microbial membrane (Meseguer et al. 1995). Halocin H7 (formerly halocin H6) from *Halpferas gibbonsii* SH7 was reported to significantly inhibit the haloarchaeal Na^+/H^+ antiporter and/or other ion transporters that span the cell membrane.

Not only can halocin H7 be used as a tool for understanding the mechanisms of the Na^+/H^+ antiporter (Meseguer et al. 1995), but it also functions as an inhibitor of the Na^+/H^+ antiporters present in eubacteria and higher eukaryotes (Na^+/H^+ exchangers). For example, halocin H7 was used to treat a dog myocardium that underwent coronary occlusion followed by reperfusion and reduced the infarct size and ectopic beats of the myocardium (Alberola et al. 1998). Lequerica and coworkers have also used halocin H7 to determine whether it could protect human skeletal cells from the myocardium against ischemia, which is the intracellular acidification and activation of Na^+/H^+ exchangers (Lequerica et al. 2006). Interestingly, in vitro experimental results showed that halocin H7 inhibited Na^+/H^+ exchangers in mammalian cells in a dose-dependent manner. The intravenous administration of this halocin into healthy mongrel dogs also exhibited protective effects against ischemia and reperfusion injury, reducing the infarct size and reperfusion arrhythmia. Thus, halocin H7 may be a lead agent for treating reperfused ischemic transplanted organs in animals. Once the mechanisms of action of other halocins are fully elucidated, these protein/peptide antibiotics may have clinical applications. In addition, the antibiotic activity of these proteins/peptides could potentially be used as archaeocin-resistant markers.

In addition to bioactive protein/peptides, other halophiles have been identified to produce bioactive secondary metabolites. In 1995, Fu and coworkers isolated two new nitrotyramine derivatives, *N*-(2-methylpropionyl)-3-nitrotyramine **112** and *N*-3(-methylbutanoyl)-3-nitrotyramine **113** (Fig. 8), from an anaerobic, halophilic eubacterium collected from the Great Salt Plains in Oklahoma, USA (Fu et al. 1995). Intriguingly, nitro compounds are not commonly isolated from microorganisms, especially anaerobes. The structural difference between compounds **112** and **113** is the additional methylene in the alkyl ester side chain. Surprisingly, even with such a minor structural difference, only compound **113** exhibited cytotoxic activity against murine leukemia P-388 cells with an IC_{50} value of 11 μM.

Other cytotoxic agents have been isolated from halophiles. For example, in 2007, two novel cytotoxic quinone-type metabolites, variecolorquinones A and B (**114–115**) (Fig. 8), were isolated from the halotolerant fungus *Aspergillus variecolor* B-17 by Wang et al. (2007a). This fungus was isolated from sediments collected from the Jilantai salt field in Alashan, Inner Mongolia, China.

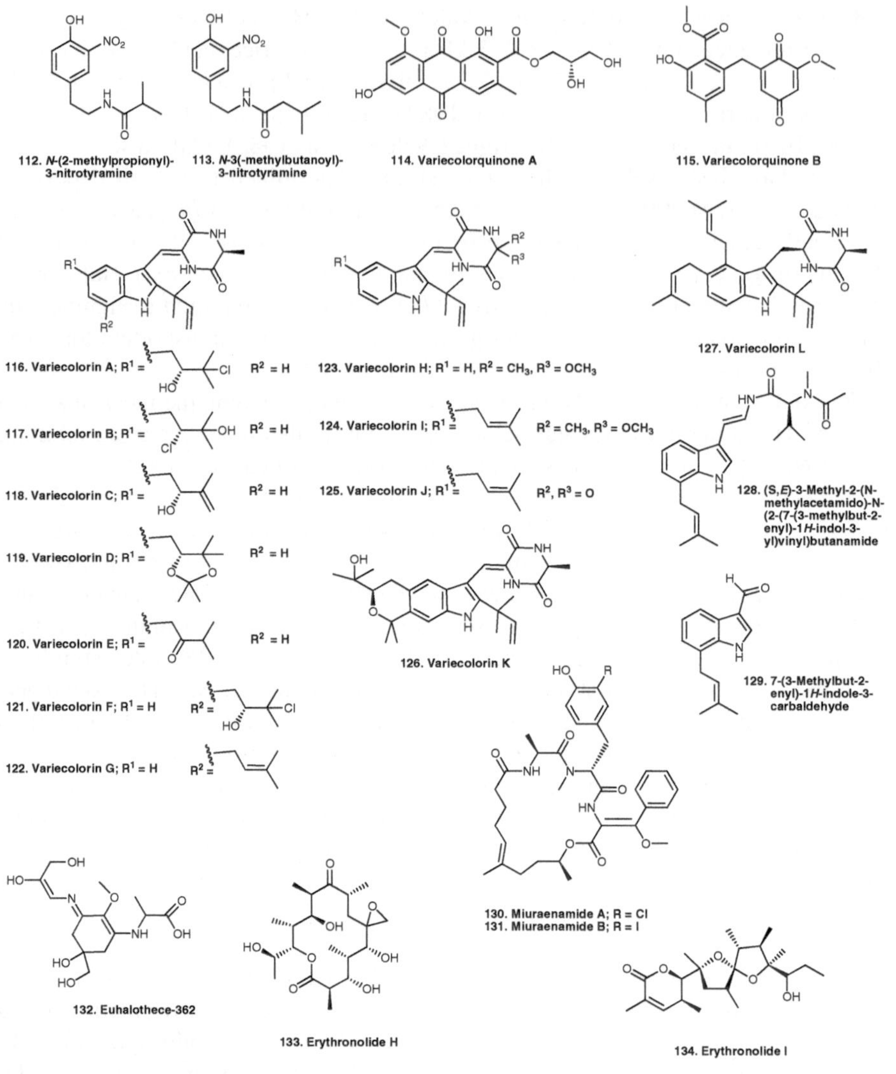

Fig. 8 Structures **112–134**

Variecolorquinone A **114** possesses an anthraquinone core unlike the structure of variecolorquinone B **115**, which is composed of benzoate and benzoquinone rings linked via a methylene group. Variecolorquinones A and B (**114–115**) exhibited moderate to weak cytotoxic activity against murine leukemia P-388 cells as well as human promyelocytic leukemia HL-60, hepatocellular carcinoma BEL-7402, and lung adenocarcinoma A549 cell lines with ranging IC_{50} values of 3–309 and 1.3–56 μM, respectively. The structural differences between **114** and **115** appeared to have an effect on the biological activity of these compounds, as variecolorquinone B **115** exhibited more potent cytotoxic activity with lower IC_{50} values.

Variecolorquinone A **114** selectively inhibited the proliferation of the human lung adenocarcinoma A549 cells, while variecolorquinone B **115** selectively inhibited the proliferation of murine leukemia P-388 and human promyelocytic leukemia HL-60 cells. Unlike variecolorquinone B **115,** compound **114** exhibited weak radical scavenging activity against DPPH with an IC_{50} value of 28 μM, which was similar to that of the ascorbic acid positive control (IC_{50}, 22 μM). These results are not surprising based on the structural differences between these compounds.

The same research group isolated 12 new isoechilin-type alkaloids, variecolorins A–L (**116–127**) (Fig. 8), from the same strain of *A. variecolor* obtained from the same site (Wang et al. 2007b). The variecolorins are essentially composed of an indole, a 2-methyl-3-buten-2-yl group, and a dipiperazine moiety with modifications to the indole and/or dipiperazine rings. Variecolorins A–K (**116–126**) exhibited weak radical scavenging activity with IC_{50} values ranging from 43 to 104 μM, and variecolorins A–L (**116–127**) were determined to be noncytotoxic. Recently, Wang and coworkers isolated a new cytotoxic indole-3-ethenamide **128** as well as other known compounds, including 7-(3-methylbut-2-enyl)-1*H*-indole-3-carbaldehyde **129**, which had never been isolated from a natural source (Fig. 8) (Wang et al. 2011c). These compounds were isolated from the halotolerant fungus *A. sclerotiorum* PTO6-1 obtained from the sediments of the Putian salt field, Fujian Province of China. Compound **128** was determined to be (*S,E*)-3-methyl-2-(*N*-methylacetamido)-*N*-(2-(7-(3-methylbut-2-enyl)-1*H*-indol-3-yl) vinyl)butanamide, from which compound **129** is proposed to be an off-pathway derivative. Compound **128** exhibited cytotoxic activity against human lung adenocarcinoma A549 and promyelocytic leukemia HL-60 cells with IC_{50} values of 3.0 and 27 μM, respectively. Although potent radical scavenging and cytotoxic activity were not observed for all of the new compounds isolated by Wang et al., these studies are examples of the structural and biological diversity that exists among halophiles obtained from soil (Wang et al. 2011c).

Potent antimicrobial agents have also been isolated from halophiles. A slightly halophilic myxobacterial strain SMH-27-4 collected from a soil sample obtained in a brush vegetation near the seashore of Arai-Hama beach in Kanagawa, Japan, was determined to produce two antibiotic depsipeptides, miuraenamides A and B (**130–131**) (Iizuka et al. 2006). Both metabolites inhibited NADH oxidase with IC_{50} values of 50 μM, suggesting that they target the electron transfer system of the mitochondrial respiratory chain. Using a paper diffusion assay, these compounds were determined to also inhibit the growth of the phytopathogenic oomycete, *Phytophthora capsici*, at 25 ng/disk. Miuraenamide A **130** exhibited potent inhibition against the phytopathogen *Phytophthora* sp. (MIC, 0.4 μg/mL), moderate inhibition against fungi and yeast (MIC values ranging from 6.3 to 12.5 μg/mL), and no inhibition against eubacteria (MIC, >50 μg/ml). During this same year, another new mycosporine-like amino acid, euhalothece-362 **132**, was isolated from the unicellular, halophilic cyanobacterium *Euhalothece* sp. strain LK-1 (Volkmann et al. 2006). Mycosporine-like amino acids contain either cyclohexenone or cyclohexeneimine chromophores conjugated with one or two amino acids. These compounds have been reported to be UV protectants (Karentz et al. 1991).

Notably, several mycosporine-like compounds have been isolated from halophilic cyanobacteria to protect these microorganisms from damaging radiation (Sinha and Häder 2008).

Novel congeners of the clinically used antibacterial erythromycin A originally developed by Abbott, erythronolides H and I (**133–134**), were isolated from the halophilic actinomycetes *Actinopolyspora* sp. YIM90600 collected by the Shen group from a dried salt lake in Xingjiang Province, China (Huang et al. 2009). Culture extracts of *Actinopolyspora* sp. YIM90600 were screened and determined to exhibit significant antibiotic activity and moderate cytotoxicity. The active components, erythronolides H and I (**133–134**), were isolated and determined to have a unique epoxide and two lactone rings, respectively. Although the authors did not report biological activity for the purified erythromycin congeners, this halophile is the first strain belonging to the genus *Actinopolyspora* that produces bioactive natural products. In previous studies, erythromycin derivatives have exhibited antibiotic activity as well as inhibit HIV-1 replication in macrophages by modulating kinase activity (Komuro et al. 2008). Based on the success of several third-generation erythromycin antibiotics in the clinic, efforts to diversify the erythromycin A scaffold in order to minimize side effects have constituted a major area of research. Thus, actinomycetes of the genus *Actinopolysora* should be further investigated for the production of other bioactive derivatives of erythromycin A.

Several screens of halophiles have been conducted that have not necessarily lead to the isolation of new active agents, but some have provided enough material to determine the bioactivity of previously reported natural products (Tian et al. 2013), demonstrating the potential of halophiles to be renewable sources of useful metabolites. For example, using a eubacterial antagonism assay, Socha and coworkers isolated the halophilic *B. endophyticus* collected from a microbial mat on the eastern shore of Salt Pond, San Salvador Island, Bahamas, that produced the known bacillamide A (Jeong et al. 2003) and two new tryptamide thiazoles, bacillamides B and C (**135–136**) (Fig. 9) (Socha et al. 2007). Almost 30 years before, bacillamide D **137** (Fig. 9), previously referred to as TM-64 in Sect. 2 on thermophiles, was isolated (Ōmura et al. 1975; Churro et al. 2009). Bacillamide A, tryptamine linked to 2-acetylthiazole-4-carboxylic acid, was previously determined to exhibit antibiosis activity against cyanobacteria, dinoflagellates (*Cochlodinium polykrikoides*, LC$_{50}$, 3.2 µg/ml), and raphidophytes (Jeong et al. 2003). However, bacillamides B–C (**135–136**) were not tested for algicidal activity due to insufficient material. None of the bacillamides exhibited antibacterial activity against *Bacillus* sp. obtained from the hypersaline pond at concentrations below 500 µM. Therefore, bacillamides B–C (**135–136**) do not appear to be the active component for the antibiosis initially observed in the screening of crude extracts. Nevertheless, *B. endophyticus* is a new source of bacillamides.

Structurally novel citrinum dimers have also been reported from a halotolerant fungal strain *Penicillium citrinum* B-57 isolated from sediments in the Jilantai salt field in Inner Mongolia, China (Lu et al. 2008). Pennicitrinone C **138** and penicitrinol B **139** were isolated and determined to be citrinin dimers (Fig. 9).

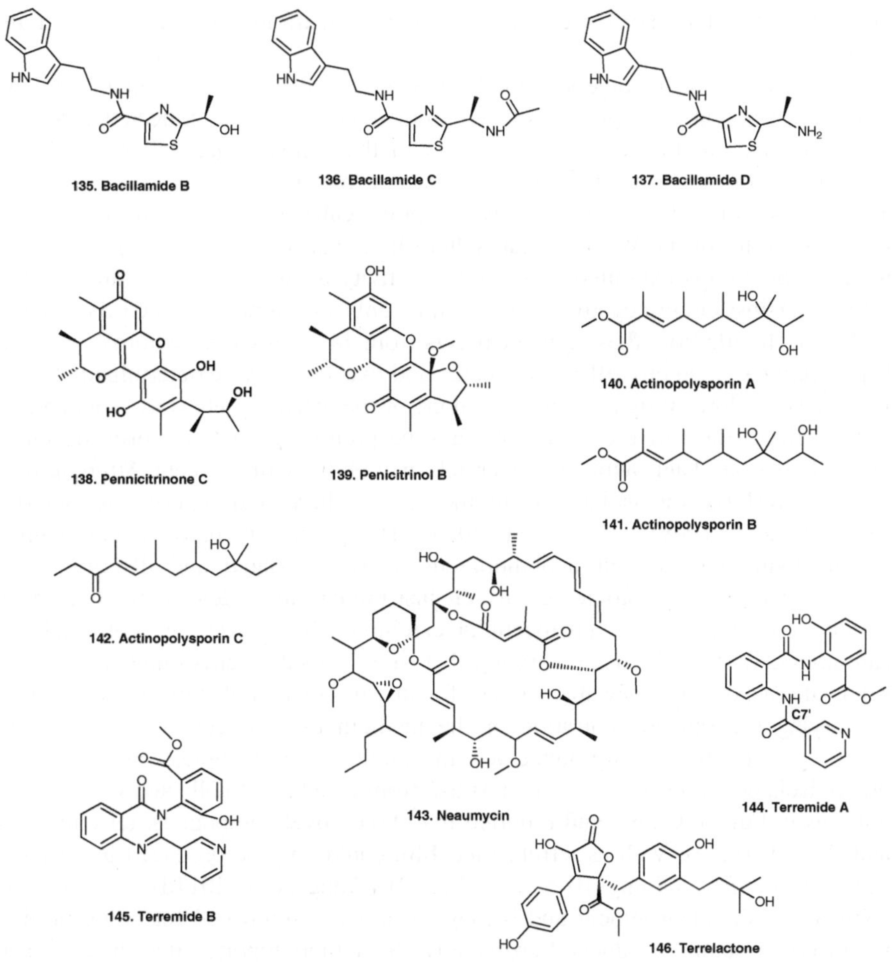

135. Bacillamide B

136. Bacillamide C

137. Bacillamide D

138. Pennicitrinone C

139. Penicitrinol B

140. Actinopolysporin A

141. Actinopolysporin B

142. Actinopolysporin C

143. Neaumycin

144. Terremide A

145. Terremide B

146. Terrelactone

Fig. 9 Structures **135–146**

Compound **138** is a pentacyclic molecule possessing a pyrone ring as well as a lactone ring with a methoxy group attached, while pennicitrinone C **138** possesses an open-ring lactone without a methoxy group. Interestingly, compound **138** exhibited moderate antioxidant activity against DPPH radicals compared to that of the ascorbic acid control (IC$_{50}$, 22 μM) with an IC$_{50}$ value of 55.3 μM. Several other reports on either the bioactivity and/or isolation of novel, structurally complex molecules from halophiles have been published (Meklat et al. 2011; Thumar et al. 2010; Gesheva and Vasileva-Tonkova 2012; Zhao et al. 2011; Tian et al. 2013; Huang et al. 2012; Sepcic et al. 2010). Some molecules, such as actinopolysporins A–C (**140–142**) (Zhao et al. 2011) or neaumycin **143** (Huang et al. 2012), isolated from *Actinopolyspora erythraea* YIM 90600 and *Streptomyces* sp. NEAU-x211, respectively, have either little to no bioactivity or simply have not been tested, but may serve as useful chemical scaffolds to develop bioactivity (Fig. 9).

There is significant diversity that exists among halophiles isolated from the same environment. Halophiles isolated from microorganisms collected from the same site, such as the sediment and water of the Weihai Solar Saltern in China, have led to the identification of new bioactive metabolites. For example, Chen and coworkers isolated 45 moderately halophilic eubacterial strains from the sediment and water of the Weihai Solar Saltern in China (Chen et al. 2010). Twenty-three of the strains exhibited antibacterial activity against *B. subtilis* and weaker activity against Gram-negative eubacteria, as only one strain inhibited the growth of *E. coli*. In addition, these crude extracts from 14 strains were cytotoxic against hepatocellular carcinoma BEL-7402 cells, and five strains were determined to have IC_{50} values below 40 μg/ml. Although none of the active agents were isolated and characterized, the active components may be promising leads for drug development. Two years later, another moderately halophilic actinomycete, *Streptomyces* sp. nov. WH26, was isolated from the same saltern and reported to demonstrate cytotoxic activity (Liu et al. 2013). The previously reported compounds 8-*O*-methyltetrangulol and naphthomycin A were isolated and determined to exhibit micromolar cytotoxic activity against human lung adenocarcinoma A549, cervical epithelial HeLa, hepatocellular carcinoma BEL-7402, and colon adenocarcinoma HT-29 cells, demonstrating that saturated saline environments, such as solar salterns, can be renewable sources for structurally complex bioactive agents.

Varying the salt concentration of the medium used to cultivate halotolerant microorganisms may also produce new metabolites. For example, Wang and coworkers isolated strain of *A. terreus* PT06-2 from a sediment collected from Putian Salterns in Fujian, China, and reported that three novel compounds, terremides A and B (**144–145**) as well as terrelactone **146,** could only be produced in a hypersaline medium containing 10 % salt (Fig. 9) (Wang et al. 2011d). *A. terreus* is abundant in the environment and belongs to a genus renowned for being prolific producers of natural products. Surprisingly, the authors reported that the chemical diversity increased when the halotolerant fungus was cultured in medium at 10 % salinity as opposed to 0 or 3 % salinity. Terremide A **144** possesses a 1,2,3-trisubstituted benzene nucleus and is a biosynthetic precursor to terremide B **145** by undergoing ring closure via the amine and amide carbonyl (C7′) carbon and dehydration. Terremides A and B (**144–145**) exhibited weak antimicrobial activity against *S. aureus* and *Enterobacter aerogenes* with MIC values of 25 and 13 μg/ml, respectively. Terrelactone A **146** is structurally different from compounds **144–145**, possessing two phenol moieties and a lactone ring; however, this compound did not exhibit any antimicrobial activity. Thus, by modifying culture growth conditions, new structurally complex chemical entities can be obtained. See the recent paper by Zheng and coworkers for another example of new antimicrobial and cytotoxic agents produced from *A. flocculosus PT05-1* also grown in a hypersaline medium (Zheng et al. 2013).

Table 5 summarizes the bioactive agents as well as their sources and optimal salt concentrations required for growth discussed in this chapter. This list will increase

Table 5 Bioactive compounds isolated from terrestrial halophiles

Halophile	Optimal salt concentration	Source	Compound	Bioactivity	References
Eubacterial strain HNGS03	8 % (w/v)	Sediment	*N*-(2-methylpropionyl)-3-nitrotyramine **112**	N/A	Fu et al. (1995)
Eubacterial strain HNGS03	8 % (w/v)	Sediment	*N*-3(-methylbutanoyl)-3-nitrotyramine **113**	Cytotoxic	Fu et al. (1995)
Aspergillus variecolor B-17	9 % (w/v)	Sediment	Variecolorquinones A and B (**114–115**)	Cytotoxic; radical scavenger (**114**)	Wang et al. (2007)
A. variecolor B-17	9 % (w/v)	Sediment	Variecolorins A–L (**116–127**)	Antioxidant	Wang et al. (2007)
A. sclerotiorum PTO6-1	8 % (w/v)	Sediment	(*S*,*E*)-3-Methyl-2-(*N*-methylacetamido)-*N*-(2-(7-(3-methylbut-2-enyl)-1*H*-indol-3-yl)vinyl)butanamide **128**	Cytotoxic	Wang et al. (2011c)
A. sclerotiorum PTO6-1	8 % (w/v)	Sediment	7-(3-Methylbut-2-enyl)-1*H*-indole-3-carbaldehyde **129**	N/A	Wang et al. (2011c)
Myxobacterial strain SMH-27-4	1 % (w/v)	Soil	Miuraenamides A and B (**130–131**)	NADH oxidase inhibitor; antimicrobial; growth inhibitor of oocyte, *Phytophthora capsici*	Iizuka et al. (2006)
Euhalothece sp. strain LK-1	5 % (w/v)	Upper layer of a gypsum crust found at the bottom of a saltern pond	Euhalothece-362 **132**	UV protectant	Volkmann et al. (2006), Karentz et al. (1991)

(continued)

Table 5 (continued)

Halophile	Optimal salt concentration	Source	Compound	Bioactivity	References
Actinopolyspora sp. YIM90600	10 % (w/v)	Dried salt lake	Erythronolides H and I (**133–134**)	N/A	Huang et al. (2009)
Bacillus endophyticus DT266	*N/A	Pond microbial mat	Bacillamides B–D (**135–137**)	N/A	Ōmura et al. (1975), Churro et al. (2009), Socha et al. (2007)
Penicillium citrinum B-57	12 % (w/v)	Sediment	Pennicitrinone C **138**	Antioxidant	Lu et al. (2008)
P. citrinum B-57	12 % (w/v)	Sediment	Penicitrinol B **139**	N/A	Lu et al. (2008)
Actinopolyspora erythraea YIM 90600	10 % (w/v)	Salt field	Actinopolysporins A–C (**140–142**)	N/A	Zhao et al. (2011)
Streptomyces sp. NEAU-x211	10 % (w/v)	Soil	Neaumycin **143**	N/A	Huang et al. (2012b)
A. terreus PT06-2	10 % (w/v)	Sediment	Terremides A and B (**144–145**)	Antimicrobial	Wang et al. (2011d)
A. terreus PT06-2	10 % (w/v)	Sediment	Terrelactone **146**	N/A	Wang et al. (2011d)

N/A information not available

* Instant ocean was used in fermentation medium

as more novel secondary metabolites are identified in microbial genomes via the amplification of genes involved in polyketide and nonribosomal peptide synthesis (Meklat et al. 2011). Genomic sequencing has already revealed that halophiles have the capacity to be prolific producers of natural products. For example, the pyrosequencing of prokaryotic DNA obtained from a saltern (13 % salt, an intermediate saline environment) in Santa Pola, Spain, revealed reduced metabolic diversity based on simplified carbon and nitrogen cycles but significant microbial diversity comprising seven different taxa in a single hypersaline environment, as approximately 57 % of the sequences were not related to any previously reported genus (Fernández et al. 2014a). Although the microbial diversity in hypersaline environments is low among archaea, genome sequencing of *Halobacterium* NRC-1 led to the identification of 972 novel genes with no corresponding homologs in gene databases.

Furthermore, new taxa of halophiles are being identified. Recently, the metagenomic sequencing of microorganisms from the surface waters of the hypersaline Lake Tyrrell in NW Victoria, Australia, has led to the identification of new haloarchaea taxa (Narasingarao et al. 2012). Not only will more available genomes lead to the identification of gene clusters dedicated to producing unusual compounds, but scientists have also found that as the distance between hypersaline environments increases, so does the divergence within microbial flora (Ma et al. 2010; Bolhuis and Stal 2011; Fernández et al. 2014b). These observations are consistent with changing environmental factors, such as pH, temperature, and salt concentration, having pronounced affects on the composition of microbial flora. Importantly, once more novel halophiles are isolated and cultured, more bioactive pharmacophores will be isolated their unexplored metabolomes.

7 Terrestrial Extremophiles Living in Mutualistic Environments?

To date, most bioactive secondary metabolites have been isolated from terrestrial habitats, although an increasing number are now being identified from marine environments. However, within the past 2 decades, there have been more reports on valuable therapeutic secondary metabolites being produced by microorganisms inhabiting mutualistic environments. Mutualism is a common term for mutually beneficial relationships between different species. These interactions are required for defense, competition, communication, host signaling, and nutrient acquisition. For the purpose of this book, microorganisms that inhabit mutualistic environments (e.g., endosymbionts), in plants, insects, and other arthropods that are either uncultivable or unable to produce specific secondary metabolites outside of these environments are considered to be extremophiles. These microorganisms are biochemically diverse due to being sensitive to environmental changes, evolutionary changes, the possibility of horizontal gene transfers, and the influences they have on their hosts. Mutualistic niches have not been thoroughly exploited and have the potential to serve as ubiquitous sources of bioactive metabolites based on their dynamic diversity. However, by the end of this chapter, one may question whether these interactions are simply mutualistic or simply epigenetic manipulations of microorganisms inhabiting the same niche.

Endosymbionts are mutualistic microorganisms that live within the bodies or cells of another organism. For example, plant endophytes are symbionts that inhabit the living tissues of plants without having deleterious effects. For every 300,000 plant species on Earth, each plant serves as a host to one or more endophytes (Strobel and Daisy 2003). Thus, one can only imagine how abundant and capable these symbionts are to produce a myriad of novel secondary metabolites. Within the past 2 decades, there have been increasing reports of only culturable endophytes producing clinically used natural products. A significant number of reports from the People's Republic of China have revealed that endophytic fungi produce the following four major classes of "plant-derived" natural products: taxanes, podophyllotoxins, camptothecins, and vinca alkaloids. For a summary of the known endophytic fungi that produce these plant secondary metabolites, see the 2012 and 2013 reviews by Chandra (2012) and Kusari et al. (2012, 2013). Culturable endophytic actinobacteria, albeit limited, have also enhanced metabolite yields (Tiwari et al. 2010) and produce new compounds (Strobel et al. 2004; Zheng et al. 2008). In 2007, Lu and Shen reported a new cytotoxic ansamycin, napthomycin K, produced by the endophytic *Streptomyces sp. CS* isolated from the medicinal plant *Maytenus hookeri* (Lu and Shen 2007). More recently, Igarashi and coworkers have identified a new anthraquinone, lupinacidin C, from the endophytic actinomycete, *Micromonospora lupine*, coexisting in the root nodules of the legume *Lupinus angustifolius*, which exhibits anti-invasive activity against murine colon cancer cells (Igarashi et al. 2011). These examples highlight the exciting new possibilities of only culturable endophytes. However, there is a mélange of other symbionts to be explored that are uncultivable outside of their mutualistic niches and depend on other organisms to produce secondary metabolites.

Approximately >99 % of environmental microbes are unculturable under standard laboratory conditions. In order to culture these microorganisms, an in-depth understanding of the ecosystem they inhabit is essential. There have been a few successful examples of how understanding mutualistic interactions has led to the development of fermentation methods to produce lead compounds. One interesting example is the production of 3′deoxyadenosine or cordycepin **147** (Fig. 10) from the unculturable entomopathogenic fungus *Cordyceps militaris* (Cunningham et al. 1950; Bentley et al. 1951). Cordycepin **147** has been used ethnopharmacologically to treat a broad range of diseases, such as rheumatism and diarrhea. As a result, *C. militaris* is often referred to as "Himalayan Viagra" or "Himalayan Gold." The main active metabolite cordycepin **147** alone exhibits potent antitumor, antimicrobial, anti-inflammatory, antimalarial, and antioxidant activity and has other useful biological properties, which are summarized in the 2014 review by Tuli et al. (2014). Cordycepin **147** has been reported to inhibit purine biosynthesis, terminates transcription, and mTOR signaling pathways.

Since its identification, several syntheses of cordycepin **147** have been published (Ito et al. 1981; Hansske and Robins 1985; McDonald and Gleason 1995); however,

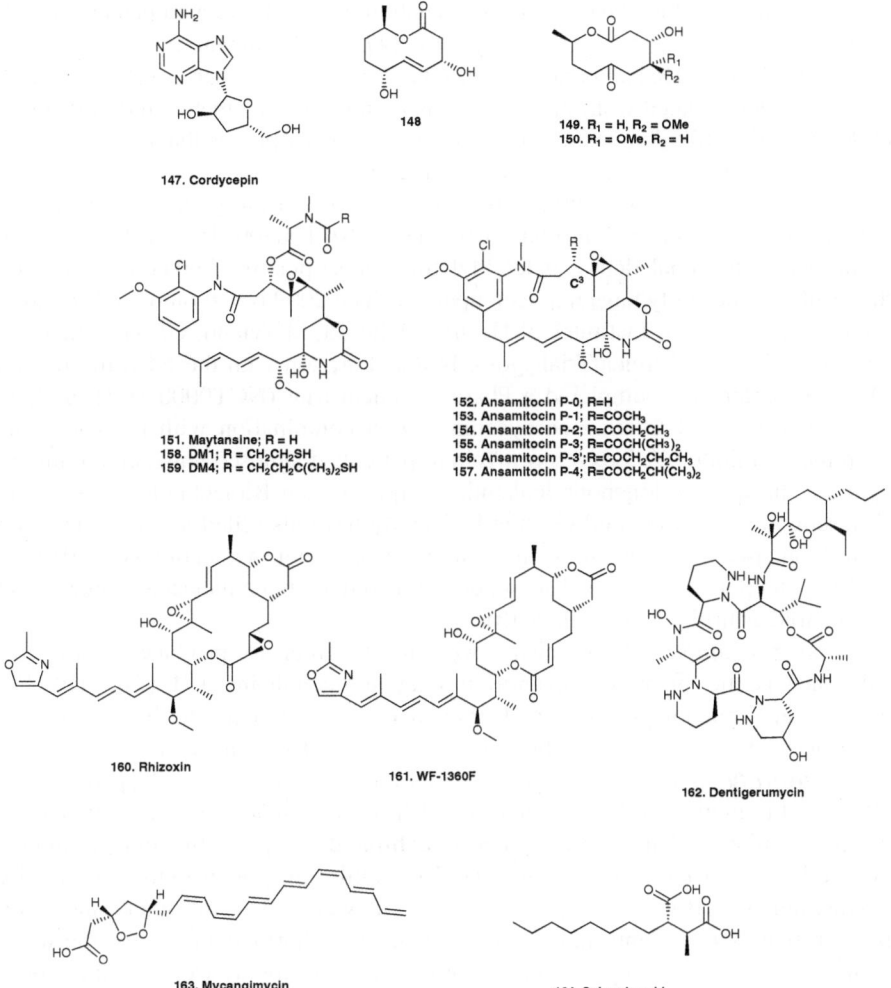

Fig. 10 Structures **147–164**

when scaling up its synthesis, the yields decrease and the reactions require several steps that utilize a large volume of environmentally unfriendly solvents (Aman et al. 2000). To obtain compound **147** from nature can also be quite challenging. Cordycepin **147** is mainly produced in the *C. militaris* fruiting body (0.97 %), relative to the corpus (0.36 %), and specific conditions are required to produce the very small blade-like fruiting bodies, which are only produced annually. During the winter, *C. militaris* invades lepidopteran larvae and the pupae of insects or spiders, on which it forms fruiting bodies in the summer. To obtain more of this mushroom, researchers have determined that the mycelium biomass has to be artificially cultivated on insect larvae or cereal to produce fruiting bodies, which contain 80–85 %

of the mushroom's bioactive secondary metabolites. There are also patents on culturing, extracting, and purifying cordycepin **147** from a mutant caterpillar fungus strain NBRC 9787 (Sakakibara and Masuda 2009, 2013). Other novel compounds have also been isolated from this fungus, including three 10-membered macrolides (**148–150**) (Fig. 10), demonstrating the diverse pharmacophores that can be isolated from uncultivable microorganisms (Paterson 2008).

Based on its cytotoxic activity, cordycepin **147** is currently under active development at OncoVista, Inc., in Phase I/II studies (NCT00709215) for the treatment of refractory terminal deoxynucleotidyl transferase-positive leukemia. However, the results of the study have not been updated since 2009 based on the information on the NIH (National Institutes of Health, Bethesda, Maryland, USA) clinical trials database (http://ClinicalTrials.gov). In collaboration with the NCI, the Boston Medical Center also completed a Phase I clinical trial (NCT00003005) to study the effectiveness of treating cordycepin **147** in combination with the adenosine deaminase inhibitor 2'-deoxycoformycin to patients diagnosed with acute lymphocytic or chronic myelogenous leukemia. Beijin Gragen Biotechnology Co., Ltd., Gingseng Science, Inc., and Gachon University have also filed several patents on cordycepin **147** for its use in formulating antidepressant medication (Du 2014) as well as osteoporosis, Paget's disease, dental disorders, bone metastatic cancer, and rheumatoid arthritis (Kim et al. 2013).

Another example of organisms working together to produce a bioactive substance is the story of the antitumor agent maytansine **151** (Fig. 10). For many years, maytansine **151** and congeners were thought to be exclusively plant-derived secondary metabolites. Maytansine **151** was isolated from the *Maytenusovatus* plant collected in Ethiopia and first reported by Kupchen et al. (1972). This metabolite was later isolated from *M. buchananii* and *Putterlickia verrucosa* plants. This compound also exhibited antiparasitic and antimicrobial activity, and based on maytansine **151** exhibiting potent cytotoxic activity against human KB cells (ED$_{50}$, 14–140 μM) as well as several other cancer cell lines, researchers became interested in using this pharmacophore for the treatment of cancer. The Meyers and Corey research groups as well as others published several papers on the total synthesis of maytansine **151** (Corey et al. 1980; Meyers and Shaw 1974); however, these syntheses were multistep, time intensive, labor intensive, and impractical for large-scale synthesis for clinical trials. To obtain enough material for clinical trials, maytansine **151** was extracted from kilograms of dried *M. buchananii* bark and stems (Kupchan et al. 1977). To generate more of compound **151** via cell culture, researchers sought out its natural source. However, being a 19-membered, halogenated ansamycin, an unusual structure for a plant secondary metabolite, that is commonly produced by eubacteria and present in some but not all individual *P. verrucosa* plants, a search commenced for microorganisms (fungal or eubacterial endophytes) that could produce its core structure.

In 1977, investigators at Tadeka Industries in Japan reported the discovery of ansamitocins P-0, P-1, P-2, P-3, P-3', and P-4 (**152–157**) (Fig. 10), which are maytansine **151** derivatives with either an ester or hydroxyl moiety at C3, from two subspecies of *Nocardia* (subsequently renamed as *Actinosynenna pretiosum*) isolated from the *Carex* species of grassy plants (Higashide et al. 1977). Because the only difference between maytansine **151** and ansamitocin P-3 **155** is the ester moiety at C3 and none of the biosynthetic genes leading to the production of maytansine **151** had been found in the plant host (Yu et al. 2012), researchers speculated that the P-3 precursor was produced by an endophyte or symbiont in the rhizosphere that is subsequently taken up by plant and converted into maytansine. This hypothesis seemed plausible as several ansamitocins are produced by eubacteria, higher plants, and mosses, contradicting the common evolutionary theory that natural products are produced by taxonomically related organisms. Wings and coworkers grew axenic cultures of *P. verrucosa* and could not amplify genes involved in maytansine **151** biosynthesis, and a maytansine-producing eubacterium could not be cultured outside of its natural habitat (Wings et al. 2013). However, using molecular techniques, such as rDNA sequencing and single-strand conformation polymorphism, they identified that the *A. pretiosum* ssp. *auranticum* eubacterium present in the rhizosphere of the plant is involved in maytansine **151** biosynthesis.

Based on rDNA sequence analysis, the *A. pretiosum* ssp. *auranticum* eubacterium had the identical 16S rDNA sequence as that amplified from the DNA of a maytansine-producing *P. verrucosa* plant (Wings et al. 2013). Other nonmaytansine-producing *P. verrucosa* plants lacked this 16S rDNA sequence. These data are consistent with the absence of maytansine **151** in cell cultures derived from maytansine-producing *P. verrucosa* plants as well as greenhouse grown *Maytenus* sp. and *Putterlickia* sp. plants and their corresponding cell cultures (Yu et al. 2012). Growing evidence has shown that the microorganisms in the rhizosphere of plants in different environments as well as those in nonrhizosphere communities in the surrounding soil appear to differ (Gunatilaka 2006), which explains why maytansine **151** is found in mosses and higher plants. However, ansamitocin-producing plants have been speculated to contribute to the structural diversity of ansamitocins via infection of the root system because only two known ansamitocins have been found in eubacteria, while there are 22 known ansamitocins in plants (Wings et al. 2013). In addition, among the 20 known eubacterial ansamitocins, eubacteria only produce 18 (Cassady et al. 2004). This is not surprising because plants survive pathogenic infection by constantly evolving their chemical defenses. Thus, plant hosts or other symbionts may be involved in the diversification of ansamitocins, a class of compounds that have served as pharmaceutical leads.

Maytansine **151** entered clinical trials in 1975 at the NCI, and after evaluation against a total of 36 different tumors in 819 patients, it was removed from trials in 1984. However, all was not lost because in the early 2000s, slightly modified structures related to ansamitocin P-3 **155** (DM1 **158** and DM4 **159**) have been linked to monoclonal antibodies directed toward tumor-linked epitopes (Alley et al. 2010). For example, a biologics license application (BLA) for trastuzumab emtansine, DM1 conjugated to Herceptin® via a thioether bridge, was submitted in August 2012 as well as a submission to the FDA for approval to be used to treat breast cancer patients (Kümler et al. 2011). On February 22, 2013, the FDA-approved ado-trastuzumab emtansine (KADCYLA®) to treat patients with

HER2-positive, metastatic breast cancer who have been previously treated with trastuzumab, a taxane, or a combination of both. The basis for studying these compounds is discussed at length in the 2010 review by Lambert of ImmunoGen, the originators of these agents and linkers, and should be consulted for further information (Lambert 2010).

In addition to trastuzumab emtansine, there are two conjugates that utilize the same DM1 **158** (Fig. 10) warhead but different monoclonal antibody carriers. Lorvotuzumab mertansine, possessing a monoclonal antibody against the CD56 epitope via a stable disulfide, is in Phase II trials in combination with lenalidomide and dexamethasone against multiple myelomas under the aegis of ImmunoGen with the antibody being provided by GTC Biotherapeutics (Berdeja 2013). Immunogen also has a variation of **158**, K7153A-SMCC-DM1, with a noncleavable thiolinker joining ansamitocin DM1 **158** and humanized IgG1 antibody K7153A (targets CD37 epitope), in a Phase I trial (NCT01534715) for the treatment of non-Hodgkin's lymphoma. The closely related DM4 **159** (Fig. 10) has also been linked to monoclonal antibodies, and there are currently four variations in clinical trials. M9346A-sulfo-SPDB-DM4, in which DM4 **159** is linked to a FOLR-1 directed MAb (folate 1 receptor) using a cleavable sulfo-SPDB linkage, is being evaluated in Phase I studies (NCT01609556) for the treatment of solid tumors under the auspices of ImmunoGen. The conjugate huDS6-DM4, in which the DS6 MAb targets the *Muc-1* epitope is also in Phase I trials sponsored by Sanofi for the treatment of solid tumors (NCT01156870). In addition, the conjugate nBT062-SPDB-DM4, in which the MAb is directed against the CD138 epitope and has a hindered disulfide linker, is in Phase I/II studies for the treatment of relapsed multiple myeloma (NCT01001442) as well as Phase I/II trials (NCT01638936) in combination with lemalidomide and dexamethasone sponsored by Biotest Pharmaceutical Corporation. Finally, huB4-DM4, in which the MAb is directed against the CD19 epitope, is in Phase II trials (NCT01472887) for the treatment of diffuse large B cell lymphoma. Even though maytansine **151** itself was abandoned in the mid-1980s, old toxic compounds, including ansamitocin derivatives, are making a comeback as directed warheads and have been approved to be safe for clinical use. Thus, mutualistic associations have the potential to be extremely useful in the search for lead compounds for drug development.

Another example of a potential antitumor drug that is biosynthesized via mutualistic interactions is rhizoxin **160** (Fig. 10). This compound is a 16-membered macrolide with an oxazole moiety originally reported to be produced by *Rhizopus* sp., a phytopathogenic fungus that causes rice seedling (*Oryza sativa* host) blight, which is the abnormal swelling of rice seedling roots (Iwasaki et al. 1984). Rhizoxin **160** inhibits rice cell division by binding to rice β-tubulin, inhibiting cell division. Furthermore, compound **160** exhibited potent antifungal activity against 10 different fungi with MIC values <1 μg/ml as well as antitumor activity in vitro and in vivo against human and murine vincristine-resistant tumor cells (Tsuruo et al. 1986). Rhizoxin **160** exhibited nanomolar cytotoxic activity against murine leukemia P-388 cells with an IC_{50} value of 91 nM. This compound was determined to bind to β-tubulin at the same site as maytansine **151**, which possesses many structural similarities to rhizoxin **160**, inhibiting tubulin polymerization and stalling mitosis, possibly via the same mechanism (Hamel 1992). Because rhizoxin **160** increased the life span (a maximum of 60 %) of mice inoculated with P-388 leukemia vincristine-resistant tumor cells and exhibited increased cytotoxicity in these cells compared to that of vincristine, it was considered to be a lead compound for developing a new chemotherapeutic agent.

Many chemists have used total synthesis to make rhizoxin **160** and several derivatives (Nakada et al. 1993; Hong and White 2004). However, to biosynthesize the antimitotic agent via the fermentation of *Rhizopus* sp., Partida-Martinez and Hertweck determined that rhizoxin **160** was not a fungal metabolite but rather a product of the eubacterial endosymbiont *Burkholderia* sp. This eubacterium contains biosynthetic genes involved in the production of rhizoxin **160** (Partida-Martinez and Hertweck 2005, 2007). These observations were consistent with four *Rhizopus* species producing rhizoxin **160** and two species that did not, all of which were obtained from distant geographical areas. Furthermore, laser microscopic observations of *Rhizopus* sp. mycelium stained with a mixture of bacteria-specific dyes revealed the appearance of a high number of live endosymbiotic eubacteria within fungal cells. Notably, when *Rhizopus* sp. was cultured in the absence of the *Burkholderia* endosymbiont, rhizoxin **160** was not produced. However, when the *Burkholderia* sp. was isolated from the fungus and cultured in the absence of *Rhizopus* sp., rhizoxin **160** and potent cytotoxic derivatives (1,000–10,000 times more active against K-562 leukemia cells) were produced (Scherlach et al. 2006). Interestingly, the isolated eubacterial endosymbiont lost its ability to produce these metabolites over time, but rhizoxin **160** increased upon the reintroduction of *Rhizopus* sp. into cultures. The authors speculated that the decrease in rhizoxin **160** was most likely due to the downregulation of its biosynthetic genes in the absence of *Rhizopus* sp.

Deletion of a *Burkholderia* p450 gene involved in rhizoxin **160** biosynthesis produced didesepoxy rhizoxin derivatives, but whether this gene is involved in catalyzing the formation of both epoxide moieties in rhizoxin **160** was unclear (Scherlach et al. 2012). The epoxidation steps were also determined to be oxygen independent. To elucidate the biosynthetic steps required to install the epoxide moieties, the authors used two different *Burkholderia–Rhizopus* associations from different regions of the world that either produced rhizoxin **160** or the monoepoxide derivative WF-1360F **161** (Fig. 10) and "switched" the symbiotic associations by cross-infecting each endosymbiotic-free *R. microporus* fungus with the endosymbiotic eubacterium of the other fungus. Interestingly, the symbiotic association that previously produced rhizoxin **160** produced WF-1360F **161**, whereas the other association produced compound **160**. Thus, these results led the authors to revise their proposed mechanism of rhizoxin **160** biosynthesis in their 2005 Nature paper. These events are most likely triggered by chemical signals that are produced via the symbiotic phytotoxin production resulting from the strain-specific association of *Burkholderia* sp. and *Rhizopus* sp., which may be further influenced by plant interactions. The vertically transmitted eubacterial intracellular symbiont of *Rhizopus* sp. delivers WF-1360F **161** to the fungus, which is actually involved in

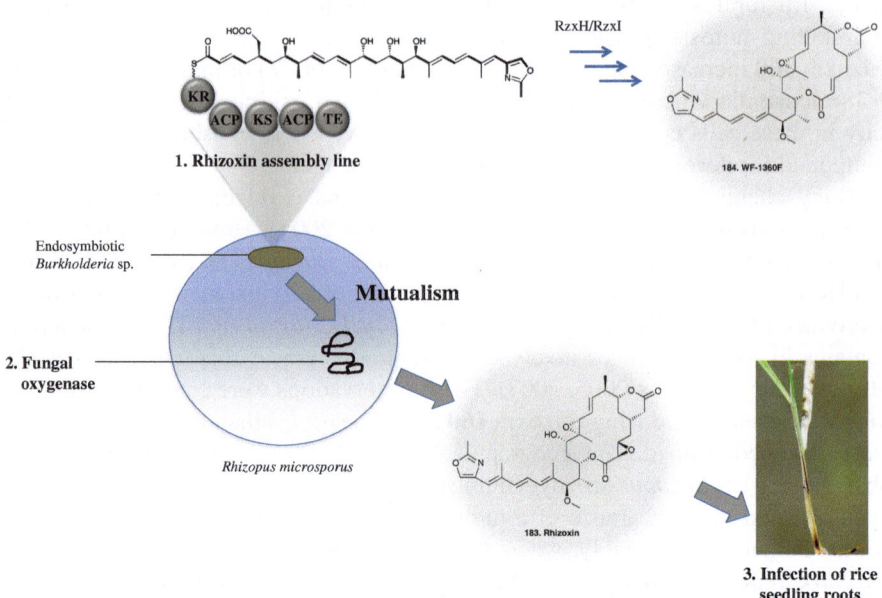

Fig. 11 Tripartite interaction between fungus and bacterium and rice seedlings

catalyzing the epoxidation of the eubacterial product **161** to make rhizoxin **160**, a more potent phytotoxin that plays an essential role in the vegetative spore formation of the fungus containing the endosymbiont, most likely for colonizing rice (Partida-Martinez et al. 2007). In this unparalleled tripartite relationship (Fig. 11), both the pathogenic fungus and endosymbiont benefit by gaining access to nutrients that are released once the phytopathogenic fungus colonizes the roots of *O. sativa*.

The NCI in the USA, EORTC in Europe, and Fujisawa in Japan selected Rhizoxin **160** for Phase I and Phase II clinical trials. In Phase I clinical trials for the treatment of refractory solid tumors, rhizoxin **160** was successfully administered as a 5-min infusion at a maximum tolerated dose of 2.6 mg/m^2 or over 72-h period via intravenous infusion at a maximum tolerated dose of 1.2 mg/m^2 (Bissett et al. 1992; Tolcher et al. 2000). Other Phase II clinical trials targeted ovarian cancer, colorectal and renal cancer, breast cancer, melanoma, squamous cell head and neck cancer, and non-small cell lung cancer. Unfortunately, the side effects observed in these studies were hematological and gastrointestinal toxicity (McLeod et al. 1996). No additional clinical studies have been reported on this compound. Perhaps, if rhizoxin **160** was a tubulin activator that did not induce peripheral neuropathy, then it would have been at the head of the queue for further drug development.

In addition to fungal–bacterial–plant symbiotic associations, interactions between microorganisms and arthropods have also led to the production of bioactive secondary metabolites that function as small-molecule control agents to ward off pathogens. These associations are pervasive throughout every ecosystem to confer survival among hosts. The seminal work published on dentigerumycin **162** (Fig. 10) by Oh and coworkers demonstrated how fungus-growing ants and actinobacteria work together to produce a specific toxin that specifically eliminates specialized fungal parasites (Oh et al. 2009a). In 2001, the eubacterium *Pseudonocardia* sp., fungal isolates, and the parasitic fungus *Escovopsis* sp. were isolated from the nest of the ant *Apterostigma dentigerum* in Gamboa, Panama. *Pseudonocardia* sp. isolated from the ant cuticle was observed to strongly inhibit *Escovopsis* sp. from the same ant colony, while the fungal isolates were resistant to this bacterium. The active component isolated from *Pseudonocardia* sp. was the depsipeptide dentigerumycin **162**, which contains highly unusual amino acid residues, such as piperazic acid, γ-hydroxypiperazic acid, β-hydroxyleucine, and *N*-hydroxyleucine, and a polyketide-derived side chain linked to a pyran ring. Dentigerumycin **162** inhibited the growth of *Escovopsis* sp. (MIC, 2.5 μg/ml), *Candida albicans* (MIC, 0.97 μg/ml), *C. albicans* ATCC10231 (MIC, 0.97 μg/ml), and amphotericin-resistant *C. albicans* ATCC200955 (MIC, 0.97 μg/ml) in liquid culture assays. Petri dish assays also showed that *Pseudonocardia* sp. inhibited the growth of *Escovopsis* sp. to a much greater extent than the fungal cultivar. Thus, the symbiosis between *Pseudonocardia* sp. and fungus-growing ants is an example of new ways ants have evolved to protect the fungal cultivar from garden parasites. Notably, the authors speculate that the eubacterial mediator *Pseudonocardia* sp. and *Escovopsis* sp. will undergo antagonistic coevolution, such that new eubacterial metabolites will target resistant *Escovopsis* sp. Such evolutionary processes will play a major role in the continuous production of new, diverse secondary metabolites from mutualistic interactions.

Analogous eubacterial mediators are common even among host–microorganisms symbioses, such as beetle–fungus mutualism. Scott and coworkers reported the existence of chemically mediated protection supplied by a eubacterial source against the fungal antagonist *Ophiostoma minus* of a fungal food source (*Entomocorticium* sp. A) required for the development of Southern pine beetle (*Dendroctonus frontalis*) larvae (Scott et al. 2008). Adult beetles harbor *Entomocorticium* sp. A in a specialized compartment, make holes in the barks of trees, deposit larvae within the bark and phloem of trees, and inoculate them with this fungus. This process can be challenged by a parasitic fungus that can outcompete *Entomocorticum* sp. A, ultimately disrupting beetle larvae development. As part of the beetle's defense mechanism, its specialized compartment harboring food is also a source of different species of actinomycetes, which are also deposited with *Entomocorticum* sp. A. The authors were able to demonstrate the antifungal activity of one actinomycete morphotype against *O. minus* with an MIC of 1.0 μM, which was 20 times more susceptible than *Entomocorticum* sp. A (MIC, 19.0 μM). The active antifungal agent was determined to be a linear 20-carbon polyunsaturated peroxide, mycangimycin **163** (Fig. 10) (Oh et al. 2009b). This compound also exhibited potent antifungal activity against *C. albicans*, *C. albicans* ATCC 10231, *C. albicans* ATCC 200955, and *Saccharomyces* *cerevisiae* at MIC values of 0.2, 0.2, 0.4, and 0.4 µg/ml, respectively. The basic scaffold of mycangimycin **163** resembles those of other antimalarial agents, and when assayed against *P. falciparum*, compound **163** exhibited antimalarial activity with an EC_{50} of 17 ng/ml, which is comparable to other antimalarial drugs with EC_{50} values close to 10 ng/ml. More studies need to be completed to determine the mechanism of action mycangimycin **163** as well as whether it possesses other biological properties. However, this is a good example of how specialized small molecules that serve as mediators within mutualistic interactions can also function as new therapeutics.

The examples of the lead compounds cordycepin **147**, maytansine **157**, rhizoxin **160**, dentigerumycin **162**, and mycangimycin **163** produced by microbes with seemingly unsustainable production without symbiotic associations or uncultivable microorganisms that grow optimally in specific niches demonstrate the potential of such "extreme environments" and evolutionary ecology to produce new pharmacophores. We know that these microorganisms exist, and most importantly, we have developed the tools to identify key biosynthetic genes in the production of these secondary metabolites, such as nonribosomal peptide synthetases and polyketide synthases, to predict whether these microorganisms can produce specific secondary metabolites. Furthermore, molecular tools, such as direct genomic DNA extraction as well as 16S and 18S rRNA and metagenomic sequencing, have revealed the presence of new taxa of eubacteria, archaea, and fungi that have not been represented by any cultured microorganism.

Metagenomic sequencing has also been used to direct culturing conditions of previously uncultivable microorganisms in communities, such as the microbiota of the medicinal leech *Hirudo verbena*, which includes *Aeromonas veronii* and a *Rikenella*-like bacterium. High expression levels of mucin and glycan utilization genes were found in the *Rikenella*-like eubacterium inside the crop, and growing the microbe in media containing mucin instead of glucose led to the growth of pure cultures of this microbe (Bomar et al. 2011). This is analagous to simulating natural environments to obtain pure cultures of microorganisms, which has also been developed as a new culturing method on a single cell level (Zengler et al. 2002). Furthermore, within the past 2 years, two papers were published back to back in Nature describing the use of 454 sequencing to identify eubacterial microbiota colonizing the root rhizosphere, soil, and endophytic compartments (within the roots) of *Arabidopsis thaliana* (Lundberg et al. 2012; Bulgarelli et al. 2012). Both studies identified similar phyla of eubacteria inhabiting the endophytic compartments of *A. thaliana* and demonstrated that they are significantly dissimilar compared to those found in plant-free soil and the root rhizosphere. Notably, the microbiota of the endophytic compartment is influenced by soil type and some variation was observed among plants of different genotypes and developmental stages. All of these observations suggest that there is a large possibility of finding more unique mutualistic microorganisms that have the capability of producing new secondary metabolites, especially when evolutionary processes occur and mutualists become resistant to chemical signals. The biogenetic basis of these metabolites is quite complex within these relationships but can lead to a range of

Table 6 Bioactive compounds isolated from terrestrial endophytes

Endophyte	Host	Compound	Bioactivity	References
Cordyceps militaris; mutant caterpillar fungus strain NBRC 9787	Lepidopteran larvae and the pupae of insects or spiders	Cordycepin **147**	Antimalarial; cytotoxic; antimicrobial; anti-inflammatory; antioxidant; antidepressant	Cunningham et al. (1950), Bentley et al. (1951), Sakakibara and Masuda (2009), Sakurai and Masuda (2013), Tuli et al. (2014), Du (2014)
C. militaris BCC 2816	N/A	10-membered macrolides (**148–150**)	N/A	Paterson (2008)
Actinosynnema pretiosum sp. *auranticum*	*Maytenusovatus; M. bucha-nanii; Putterlickia verrucosa*	Maytansine **151**	Antiparasitic; antimicrobial	Kupchan et al. (1972), Wings et al. (2013)
A. pretiosum ssp. *auranticum*	N/A	Ansamitocins P-0, P-1, P-2, P-3, P-3′, and P-4 (**152–157**)	Cytotoxic	Higashide et al. (1977)
N/A	N/A	Ansamitocin DM1 **158**	Cytotoxic	Alley et al. (2010)
N/A	N/A	Ansamitocin DM4 **159**	Cytotoxic	Alley et al. (2010)
Burkholderia sp. and *Rhizopus* sp.	*Oryza sativa*	Rhizoxin **160**	Cytotoxic; phytotoxin	Iwasaki et al. (1984), Partida-Martinez and Hertweck (2005), Partida-Martinez et al. (2007)
Burkholderia sp.	*Rhizopus* sp.	WF-1360F **161**	Antifungal; cytotoxic; phytotoxin	Loper et al. (2008)
Pseudonocardia sp.	*Apterostigma dentigerum*	Dentigerumycin **162**	Antimicrobial	Huang et al. (2009)
Entomocorticium sp. A in the presence of *Ophiiostoma minus*	*Dendroctonus frontalis*	Mycangimycin **163**	Antimalarial; antimicrobial	Scott et al. (2008), Oh et al. (2009)
Sphaeropsis sp.	*Taxus globosa*	Sphaeric acid (**164**)	Antimicrobial; affects the function of interleukin-1	Wilkinson et al. (1999)

N/A information not available

interesting signaling compounds in mutualism or antibiotics and virulence factors that result from parasitic interactions. With next-generation sequence technologies, many more reports on the sequencing of microbiomes will be published, facilitating the dissection of mutualistic interactions and providing further insight into how to possibly culture some of these organisms. In addition, the biosynthetic potential of more microorganisms that produce new chemical entities can be determined and exploited using heterologous hosts.

Lastly, the development of activating silent genes using epigenetics as well as new culturing or cocultivation methods will assist in growing uncultivable microorganisms. Some breakthroughs have been made in how to culture these organisms. For example, the sphaeric acid (**164**)-producer *Sphaeropsis* sp. can be cultured on agar plates containing small pieces of inner bark (phloem–cambium and xylem tissues) of the Mexican yew tree *Taxus globosa* (Fig. 10) (Wilkinson et al. 1999). Not only can host–symbiont interactions induce the production of new compounds, but interactions between symbionts also have the potential to produce new secondary metabolites, as hosts are unlikely to be colonized by just a single microorganism. Biosynthetic genes can be up- or downregulated in symbionts as a result of interactions with other microorganisms within their environment (Bandara et al. 2006; Nützmann et al. 2011). Signaling molecules analogous to eubacterial quorum sensing and other elicitors are thought to be involved in activating silent biosynthetic gene clusters, and several reports of coculturing microorganisms have demonstrated either the production of a new secondary metabolite or the enhanced production of an already known metabolite (Bertrand et al. 2014; Goers et al. 2014; Marmann et al. 2014).

Table 6 summarizes all of the bioactive agents solely produced by endophytic microbes discussed in this chapter. One has to wonder if the mere presence of another microbiome triggers the (over)expression of genes involved in producing a metabolite that may have already been produced in a microbe but below the detection limits of analytical instrumentation. Is this a mutualistic environment? Could this be considered an antagonistic environment that forces all microorganisms to protect themselves? Or could these interactions be viewed as being the same? For example, consider the effects of beetle–fungus mutualism without a mediator; the fungus within the host would not survive and the host would not survive because of the lack of a food source. There is also speculation about the impact of the host on the mediator, as the host houses the mediator. Nevertheless, these interactions eventually result in epigenetic adaptation, activating genes involved in producing tailored responses in the form of novel chemical entities.

8 Other Extremophiles

Microorganisms also inhabit other extreme environments, such as within, on, or underneath rocks as well as extremely dry, radioactive/toxic, metallic, and other unclassifiable environments (e.g., fecal matter). Only a few bioactive natural products have been reported from these extremophiles, most likely due to the limited number of microorganisms that have been isolated from these environments. However, as geo(microbiologist), astrobiologists, and ecologists have started to look for life beyond soils and in other habitats unseemly occupied by microorganisms, the number of papers on these environments has slowly increased and new dedicated international scientific journals have been established, such as Extremophiles, Geobiology, Geomicrobiology Journal, and Environmental Microbiology.

8.1 Endoliths

Microorganisms that avoid extreme climates by associating with porous and translucent rocks are classified as endoliths. These microbes colonize the pores of rocks and are commonly associated with a community of symbiotic microorganisms. Translucent rocks facilitate photosynthesis, which is essential for photoautotrophic energy capture for endoliths. Furthermore, endolithic environments protect microorganisms from desiccation, radiation, wind, and temperature fluctuations. The growth of endoliths is limited by the availability of CO_2 as it slowly diffuses through rocks as well as access to water via diffusion or the melting of ice or permafrost. Closely related to endoliths are euendoliths, such as *Chasmoendoliths* and *Cryptoendoliths*, which actively bore into rocks to colonize them. *Chasmoendoliths* inhabit cracks and fissures in weathering rocks and are found in ice-free areas in polar or mountainous regions where the freeze fracturing of rocks commonly occurs. *Cryptoendoliths*, typically cyanobacteria, dwell in the interstices of crystalline rock structures in desert regions worldwide. However, trained microbiologists have had difficulty distinguishing between these two types of endoliths. Molecules that possess water-retentive properties, such as trehalose and other sugars, have also been isolated from endoliths, enabling their resistance to multiple cycles of desiccation and rehydration. Several strains of *Leptolyngbya*-like cyanobacteria have been isolated in the rocks of the Mammoth region of Yellowstone National Park (Wyoming, Montana, and Idaho, USA) (Banerjee et al. 2009). Interestingly, these strains are related to *Leptolyngbya* isolates isolated from a hot spring in Greenland as well as a *Leptolyngbya* morphotype isolated from a geothermal seawater lagoon, demonstrating the potential of these organisms to produce diverse and unique chemical entities. Currently, these cyanobacteria are being exploited for the production of glycolipids, phospholipids, triacylglycerides, and other lipids, such as oleic, linoleic, and linolenic acids, for biodiesel production (Singh et al. 2014).

8.2 Other Rock-associated Microorganisms

Epiliths are microorganisms that grow on the external surfaces of rocks and commonly form microcolonies or biofilms. Although bioactive agents have not been reported from these microorganisms, some epiliths are associated with metal oxides and concentrated minerals (Dorn and Oberlander 1981) and may potentially have unique metabolomes. Hypoliths grow on the underside of translucent rocks in cold deserts and are major sources of biomass. 18S rRNA sequencing of hypolithic communities in cold and hot deserts has mostly focused on the biodiversity of their fungal inhibitats. Recently, Antarctic hypolithic communities were characterized by 18S rRNA sequencing, and several uncharacterized bryophytes, fungi, and protists were identified, some of which have the capacity to produce secondary metabolites (Gokul et al. 2013). Thus, hypoliths are most likely an unexplored source of new chemical entities. Sequencing of the 16S rRNA of other eubacterial hypolithic communities has also demonstrated the presence of significantly different profiles of cyanobacteria (e.g., predominantly of the genus *Chroococcidiopsis*) and other eubacteria, including chloroflexi and actinobacteria, which vary based on the nutrients available (Lacap et al. 2011; Wong et al. 2010). In a 2012 review, Pointing and Belnap described the biome of rock surface communities and discussed the impact the environment can have on these microorganisms, which may also lead to changes in their corresponding metabolic profiles (Pointing and Belnap 2012). A ubiquitous feature of rock-dwelling organisms appears to be the presence of cyanobacteria, which may produce unique chemical scaffolds in these extreme and dynamic environments (e.g., sodium levels, temperature, or moisture).

8.3 Xerophiles

Microorganisms that grow optimally in environments of reduced water activity and are resistant to desiccation are called xerophiles, which include many rock-dwelling organisms. Xerophiles occupy some of the world's driest regions, which may have high salinity, intense atmospheric radiation, and limited nutrients. Approximately 15 % of the Earth's terrestrial land is composed of subtropical, cool coastal, cold winter, or polar desert. Culturable microorganisms have been recovered from these environments. Eubacteria have existed in desiccating conditions for decades. For example, Vreeland and coworkers isolated the Gram-positive eubacterium *Bacillus sphaericus* from an extinct bee trapped in amber for 25–40 million years (Vreeland et al. 2000). In addition, a 250-million-year-old halotolerant bacterium, *Virgibacillus pantothenticus*, was isolated from a salt crystal from the Permian Salado Formation.

Algerian Saharan soils have been reported to harbor a significant amount of diverse actinomycetes, such as *Nocardiopsis* and *Saccharothrix* (Sabaou et al. 1998; Badji et al. 2005; Boudjella et al. 2006). These actinomycetes have been reported to produce several metabolites with antimicrobial activity. Zitouni and coworkers identified the antifungal agents ZA01 and ZA02 from a screen of 86 isolates having either *Nocardiopsis* (54 isolates) or *Saccharothrix* (32 isolates) morphology and screened their extracts for antimicrobial activity (Zitouni et al. 2005).

One strain of *Saccharothrix* sp. SA 103 exhibited antifungal activity against *Mucor ramannianus*, which is unprecedented for this genus, and antibacterial activity. The active components, ZA01 and ZA02, were purified and partially characterized to be nucleosidic antibiotics. A number of papers have been published on the identification of new bioactive agents produced by xerophiles isolated from deserts, but these molecules have not been structurally characterized, most likely due to limited access to resources to perform analytical analyses (Boudjella et al. 2006; Botić et al. 2012; Thumar et al. 2010; Hrouzek et al. 2011).

Based upon the location of the Southwest Center for Natural Products Research and Commercialization in Tucson, AZ, USA, several papers have been published on bioactive metabolites produced by xerophiles isolated from the Sonora Desert, a part of which is located in Arizona. Most of these bioactive molecules have been isolated from endophytic fungi. During a screening program initiated to identify anticancer agents from the rhizosphere microflora of Sonoran Desert plants, several fungal extracts exhibited ≥ 90 % inhibition of at least one of the following cancer lines at 10 μg/ml: non-small cell lung cancer NCI-H460, breast cancer MCF-7, and central nervous system glioma SF-286 cell lines (Wijeratne et al. 2003). A new cyclopentenedione, asterredione **165**, and two terrecyclic acid A derivatives, (+)-5(6)-dihydro-6-methoxyterrecyclic acid A **166** and (+)-5(6)-dihydro-6-hydroxyterrecyclic acid A **167**, were identified from *Aspergillus terreus* Thom growing in the rhizosphere of *Opuntia versicolor* collected from the Tucson Mountains in southern Arizona (Fig. 12). Using bioactivity-guided fractionation, asterredione **165** was isolated and determined to exhibit weak cytotoxicity against human non-small cell lung cancer NCI-H460, breast cancer MCF-7, and central nervous system glioma SF-286 cell lines with IC_{50} values of 17.4, 25.2, and 20.7 μM, respectively. This compound was recently made via total synthesis by Gau Zhang and Hu in five simple chemical steps (Gai et al. 2014). (+)-5(6)-Dihydro-6-hydroxyterrecyclic acid A **167** lacks a methoxy group (**166**) at C6. Both compounds exhibited moderate to weak cytotoxicity such that the IC_{50} values for compounds **166** and **167** ranged from 19.4–21.4 to 7.8–16.5 μM, respectively. The authors speculate that the cytotoxicity of compounds **165–167** is due to the electrophilic α-methylene carbonyl moiety covalently binding to DNA via a Michael-type reaction. Thus, these types of structures may provide insight into developing cytotoxic agents via similar functional groups to covalently modify DNA.

The same research group also identified the cytotoxic aspochalasins I–K (**168–170**) in an ethyl acetate extract of *A. flavipes*, occupying the rhizosphere of the desert turpentine brush, *Ericameria laricifolia* Nutt (Zhou et al. 2004). Aspochalasins I–K (**168–170**) possess carbocyclic [or oxygen-containing (**168–169**)] rings that connect the C8 and C9 positions of a perhydro-isoindol-1-one moiety bearing a 2-methyl-propyl group at the C3 position. These compounds structurally differ by possessing methoxy and hydroxyl groups at C17, C18, or C19 as well as a double bond between C19 and C20 (**168–169**). All compounds exhibited weak to moderate cytotoxicity against human non-small cell lung cancer NCI-H60, breast cancer MCF-7, and central nervous system glioma cancer SF-268 cell lines with IC_{50} values

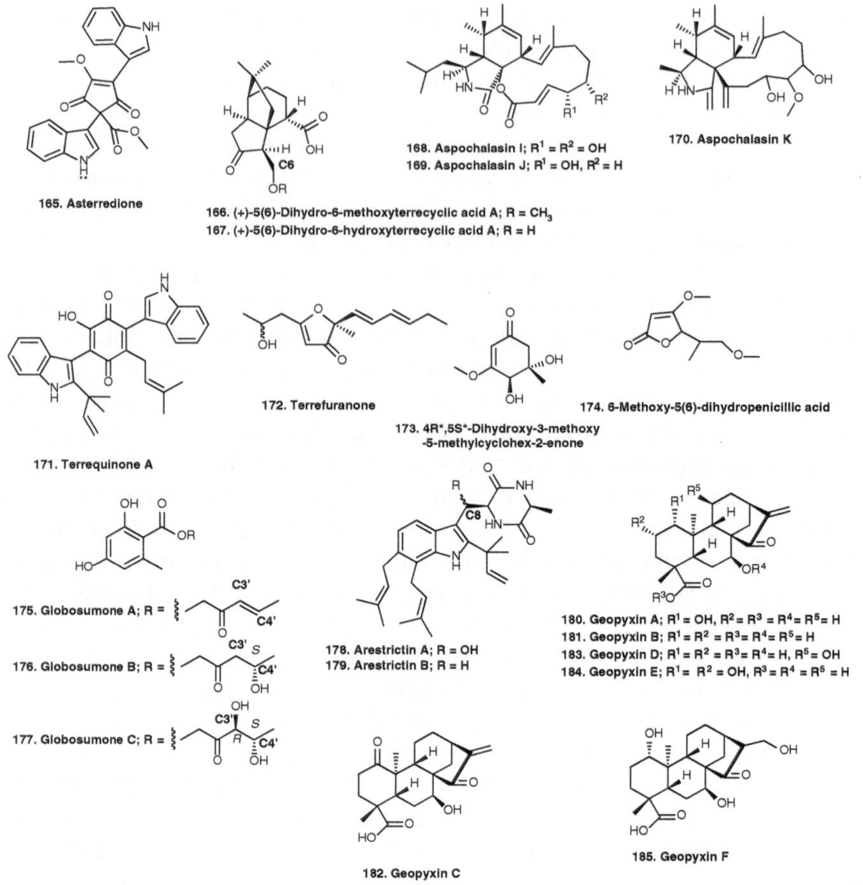

Fig. 12 Structures **165–185**

ranging from 13.1 to 55.2 μM. Based on the functional groups present in other cytotoxic aspochalasins, the authors speculate that the presence of the electrophilic α,β-unsaturated carbonyl moiety enhances cytotoxicity.

Other cytotoxic strains of *Aspergillus* have been identified from the rhizosphere of other Sonoran desert plants (He et al. 2004). The new metabolites, terrequinone A **171**, terrefuranone **172**, 4R*,5S*-dihydroxy-3-methoxy-5-methylcyclohex-2-enone **173**, and 6-methoxy-5(6)-dihydropenicillic acid **174** were isolated from these fungal strains. Compound **171** is a bisindolylbenzoquinone, whereas compound **172** is a trisubstituted dihydrofuran-4-one. However, only terrequinone A **171** exhibited moderate cytotoxic activity against human non-small cell lung cancer NCI-H460, breast cancer MCF-7, central nervous system glioma SF-268, pancreatic cancer Pa Ca-2, and normal human primary fibroblast cells with IC_{50} values of 5.60, 6.80, 13.90, and 5.40 μM, respectively. The Walsh group subsequently published the biosynthesis of compound **171**, which was elucidated by cloning its gene cluster regulated by LaeA from *A. nidulans* (Bouhired et al. 2007) and heterologously expressing the five contiguous genes in *E. coli* to evaluate the in vitro activity of the corresponding proteins (Balibar et al. 2007). Thus, there is another way of generating more terrequinone A **171** and possibly new analogs.

Novel cytotoxic orsellinic acid esters, globosumones A–C (**175–177**), were isolated by Bashyal and coworkers from the endophytic fungus *Chaetomium globosum* (Bashyal et al. 2005), which inhabited the *Ephedra fasciculata* plant obtained from the Sonoran Desert (North America). Globosumones A–C (**175–177**) have various oxygenation patterns or degrees of unsaturation across the bond between C3′ and C4′ of the pentyl side chain of the ester. Globosumones A and B (**175–176**) exhibited moderate to weak cytotoxic activity against human non-small cell lung cancer NCI-H460, breast cancer MCF-7, central nervous system glioma SF-268, pancreatic cancer MIA Pa Ca-2, and normal fibroblast WI cell lines with IC$_{50}$ values ranging from 6.50 to 30.20 µM. These results suggest that substitution at C3′ affects cytotoxicity more than the presence of a rigid double bond between C3′ and C4′.

New dioxopiperazines linked to prenylated indoles were isolated by Itabashi and coworkers from extracts of the xerophilic fungi *A. restrictus* strain A-17 and *A. pennicilliodes* SPEG. strain SUM3319 found in house dust in Hyogo prefecture, Japan, and the surface of tuna jerky collected in Yokohama city, Japan, respectively (Itabashi et al. 2006). The compounds were determined to be arestrictins A and B (**178–179**), which structurally differ by the presence of a hydroxyl group at C8 (**178**). Although the bioactivity has not been reported for these compounds, these metabolites are thought to have immunosuppressive activity based on their structural similarity to the immunosuppressive agent cristatin A.

Recently, endolichenic fungi have been reported to be sources of structurally diverse secondary bioactive metabolites (Wijeratne et al. 2012). Kithsiri Wijeratne and coworkers reported four new *ent*-kaurane diterpenoids, geopyxins A–F (**180–185**), which were isolated from the fungi *Geopyxis* aff. *Majalis* [geopyxins A–D (**180–183**)] and *Geopyxis* sp. AZ0066 (geopyxins E and F **184–185**) growing in live thallus of the intense dark and light lichen *Pseudevernia intensa* (Nyl.) Hale and W.L. Culb collected from the Sonoran Desert bioregion. Typically, fungal *ent*-kauranes are rare and have only been reported from *Gibberella fujikuroi* and *Phaeosphaeria* sp. L487. These novel *ent*-kaurene diterpenoids structurally differ based on having different hydroxylation patterns. Importantly, geopyxin B **181** exhibited weak cytotoxic activity against human non-small cell lung cancer NCI-H460, central nervous system glioma SF-268, breast cancer MCF-7, metastatic prostate adenocarcinoma cancer PC-3 M, and metastatic breast adenocarcinoma MDA-MB-231 cell lines with IC$_{50}$ values of 2.25 ± 0.15 µM, 2.35 ± 0.31 µM, 4.32 ± 1.55 µM, 5.41 ± 0.91 µM, and 3.31 ± 0.44 µM, respectively. Using a heat-shock reporter cell line, geopyxin B **181** also exhibited weak heat-shock-inducing activity within the same concentration range in which it exhibited cytotoxic activity against human cancer cell lines, suggesting that heat-shock activation is not a consequence of toxicity. Therefore, aside from the α,β-unsaturated ketone carbonyl moiety, geopyxin B **181** may have a structural targeting moiety that confers selectivity and an effector motif that modulates thiol reactivity via Michael additions. This molecule could be used to probe the mechanisms involved in maintaining cell homeostasis.

8.4 Toxitolerant Extremophiles

Organisms that are able to withstand high levels of damaging agents, such as organic solvent, are considered to be toxitolerant. For example, the pathogenic fungi *Crytoccocus neoformans* and *Candida albicans* were found to be more tolerant in media containing up to roughly 50 % dimethylsulfoxide (DMSO) and acetone (Eloff et al. 2007). Radiotolerant or radioresistant microorganisms can also be considered to be toxitolerant extremophiles. These microorganisms can survive in environments exposed to high levels of radiation, energy in the form of particles or electromagnetic waves (e.g., gamma rays, X-rays, radioisotopes, and ultraviolet radiation). Radiotolerant extremophiles are typically found in environments of high elevation (e.g., mountain ranges), industrial and biomedical research facilities that use radioactive elements, and open fields that are exposed to the ultraviolet radiation provided by the sun due to the depletion of the ozone layer. Above threshold levels, radiation is lethal to living organisms, causing oxidative damage to cells and tissues. It is unclear how microorganisms, such as *Deinococcus radiodurans*, *Rubrobacter* sp., some species of *Bacillus*, and *Dunaliella bardawil*, grow in excessive ionizing radiation (up to 20 kGy of gamma radiation) and UV radiation (up to 1,000 J/m^2) (Kumar et al. 2010). The radiotolerance of *D. radiodurans* has been proposed to result from its resistance to dessication, as these microorganisms survive by secreting radiation-protective molecules and having mechanisms for the rapid repair of DNA strand breaks. Not surprisingly, the radiotolerant microbes are a major source of photo-protective pigments and antioxidants that prevent cell and tissue damage.

To date, there are several known antioxidants and antiproliferative agents, such as ectoine, melanins, pannarin, mycosporine-like amino acids, sphaerophorin, scytonemin, and bacterioruberin, that have been isolated from radioresistant extremophiles (Gabani and Singh 2013). A novel astaxanthin derivative, astaxanthin dirhamnoside **186** (Fig. 13), was isolated by Asker and coworkers from the radiotolerant eubacterium *Sphingomonas astaxanthinifaciens* strain TDMA-17 obtained from Tottori, Japan, a region known for having high radon activity (^{226}Ra, 0.60 Bq/L; ^{228}Ra, 0.41 Bg/L) (Asker et al. 2009). Astaxanthin dirhamnoside **186** is a red carotenoid that is essentially astaxanthin with two rhamnose moieties. This derivative exhibited antioxidant activity by inhibiting methyl linolate hydroperoxide production by 27 % at 1 mM, while astaxanthin and β-carotene inhibited hydroperoxide production by only 20 and 16 %, respectively. More bioactive pigments will most likely be identified from these extremophiles as reports suggest that radiotolerance increases pigment/carotenoid production (Asker et al. 2012).

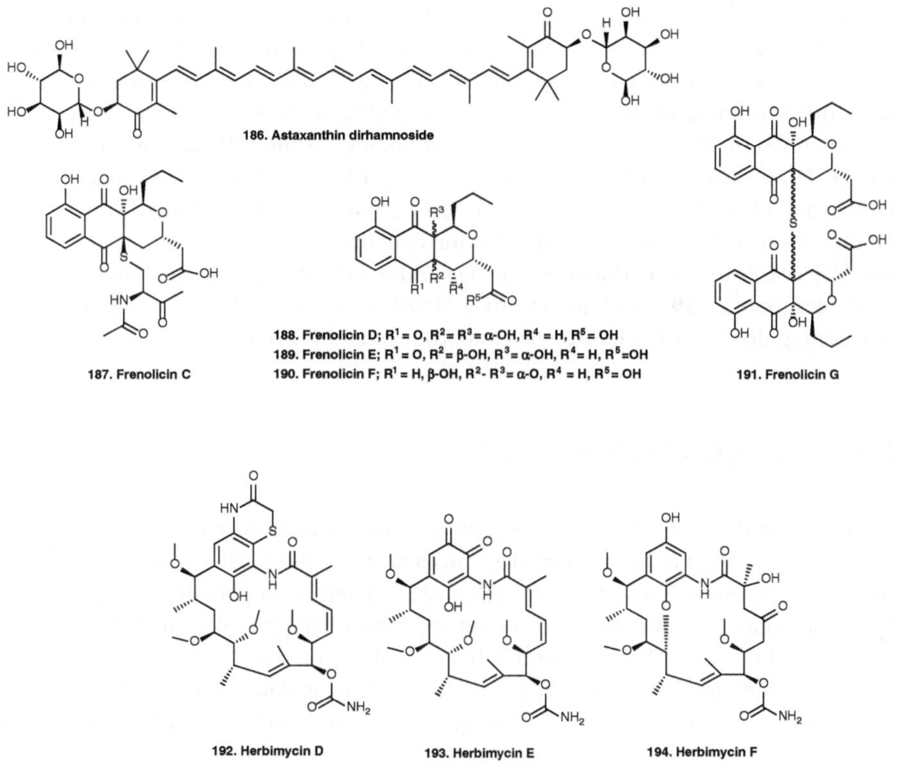

Fig. 13 Structures **186–194**

8.5 Metallotolerant Microbes

Microorganisms that are capable of surviving in environments with high levels of heavy metals, such as copper, cadmium, arsenic, and zinc, are referred to as being metallotolerant. Many of these microbes tend to exist near mines (e.g., coal) or anthropogenic environments, which are unexplored environments for new bioactive metabolites. Recently, Wang and coworkers reported the production of four new pyranonaphthoquinones, frenolicins C–G (**187–191**) (Fig. 13) from *Streptomyces* sp. RM-4-15 isolated from the man-made Appalachian Ruth Mullins underground coal fire site (Wang et al. 2013). Frenolicins C–F (**187–190**) structurally differ based upon oxygenation of the pyranonaphthoquinone core and the presence of an *N*-acetyl cysteine residue. However, when *Streptomyces* sp. RM-4-15 was grown in fermentation media containing 18 mg/L of scandium, a new homodimeric analog, frenolicin G **191** (Fig. 13), was produced. None of these new compounds exhibited any cytotoxic activity; however, this study is an example of the diverse chemical scaffolds that can be found or created using species growing in metal-enriched environments.

The same research group isolated novel ansamycin analogs, herbimycins D–F (**192–194**) (Fig. 13), from a crude extract of *Streptomyces* sp. RM-7-15, which was also obtained from a soil sample collected near a thermal vent associated with the Ruth Mullins underground coal mine fire (Shaaban et al. 2013). These compounds possess either a mercaptoacetamide moiety at the C19 position (herbimycin D, **192**), an *ortho*-quinone (herbimycin E, **193**), or an unusual ether linkage between C11 and C21 (herbimycin F, **194**). Interestingly, although these compounds were not cytotoxic against tumor cell lines, they were observed to bind the Hsp90a *N*-terminal domain with a similar affinity to that of geldanamycin. Thus, compounds **192–194** are nontoxic Hsp90 inhibitors, which are of interest for developing drugs for the treatment of neurodegenerative diseases.

8.6 Unclassifiable Extremophiles

There are other examples of extremophiles that have been isolated from unclassifiable environments and reported to produce new chemical entities. In Sect. 3, the *Microbispora aerate* strain IMBAS-11A isolated from penguin excrements on Livingston Island, Antarctica, was noted to produce the cytotoxic microbiaeratin **9**. However, there have been several other examples of novel metabolites isolated from "similar" environments. Yang and coworkers isolated a rare 20-membered macrocyclic lactam with a diene-conjugated olefin, sannastatin **195**, and known congener vicenistatin **196** from the culture broth of *S. sannanesis* obtained from the feces of the panda *Ailuropoda melanoleuca* (Fig. 14). Vicenistatin **196** is a cytotoxic 20-membered macrolactam polyketide with an amino sugar, which is a potential biosynthetic precursor to sannastatin **195** through selective epoxidation, again demonstrating how Nature synthesizes new derivatives by modifying a base structure (Yang et al. 2011). Compound **195** was determined to be toxic against brine shrimp larvae with an 82.4 % mortality rate at a concentration of 10 μg/ml. Sannastatin **195** may be a potent cytotoxic agent because the structurally related, cytotoxic chaetomugilin A control exhibited growth inhibitory activity in brine shrimp assays with a 78.3 % mortality rate at the same concentration.

Violapyrones A–G (**197–203**) (Fig. 14), a new family of 3,4,6-trisubstituted α-pyrones, were isolated by Zhang et al. from *S. violascens* YIM 100525 while screening cultivatable actinobacteria for new secondary metabolites from the feces of the five following animal species: *Hylobates hoolock*, *Rhinopithecus bieti*, *Panthera tigris altaica*, *Ailurus fulgens*, and *Viverra zibetha*. *S. violascenes* YIM 100525 was isolated from feces excreted from an adult primate (*H. hoolock*) living in Yunnan Wild Animal Park, Kunming, Yunnan Province, China (Zhang et al. 2013b). Violapyrones A–G (**197–203**) possess an α-pyrone ring with alkyl side chains of different lengths and oxygenation patterns. Compounds **197–203** were determined to exhibit antibacterial activity against *B. subtilis* and *S. aureus* with MIC values ranging from 4 to 32 μg/ml, suggesting that the oxygenation of the alkyl side chain affects bioactivity of the violapyrones. Recently, violapyrones

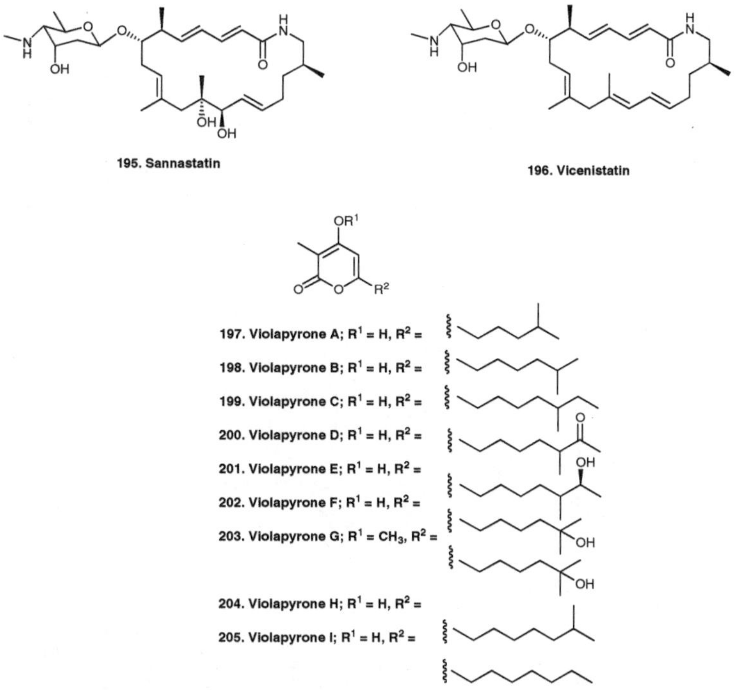

195. Sannastatin

196. Vicenistatin

197. Violapyrone A; R¹ = H, R² =
198. Violapyrone B; R¹ = H, R² =
199. Violapyrone C; R¹ = H, R² =
200. Violapyrone D; R¹ = H, R² =
201. Violapyrone E; R¹ = H, R² =
202. Violapyrone F; R¹ = H, R² =
203. Violapyrone G; R¹ = CH₃, R² =
204. Violapyrone H; R¹ = H, R² =
205. Violapyrone I; R¹ = H, R² =

Fig. 14 Structures **195–205**

H and I (**204–205**) were isolated from *Streptomyces* strain 112CH148 associated with the marine starfish *Acanthaster planci* (Shin et al. 2014). Compounds **204–205** have saturated alkyl side chains of different lengths and methylation patterns. Interestingly, violapyrones B **198**, C **199**, and H–I (**204–205**) were determined to exhibit cytotoxic activity against 10 human cancer cell lines with GI_{50} values ranging from 4.6 to 110 µM. Although these are structurally similar, the position of the methyl group in the alkyl side chain appears to modulate their bioactivity. For example, violapyrones possessing an isomethyl group were more cytotoxic. In the initial report of the isolation of violapyrones A–G (**197–203**), all compounds were observed to lack cytotoxic activity against some of the same cancer cell lines. However, this may have been due to the differences in sensitivity between the assays used to detect cytotoxic activity (MTT assay vs. sulforhodamine B assay).

Within the past decade, the number of papers published on the metagenomic analyses of the microbiome of animals, including humans, has increased significantly, as scientists try to understand the role of mutualistic microorganisms in the pathogenesis of disease (e.g., obesity and cancer), immune system development, as well as aging in the host. Metagenomics is essentially the extension of the genome sequencing approach to a collection of environmental DNA. The chemical signatures of the mammalian microflora have been shown to influence the health of the host by activating enzymes that influence metabolic pathways involved in

Table 7 Bioactive compounds isolated from other terrestrial extremophiles

Extremophile	Classification	Compound	Bioactivity	References
Aspergillus terreus Thom	Xerophile	Asterredione **165**	Weakly cytotoxic	Wijeratne et al. (2003)
A. terreus Thom	Xerophile	(+)-5(6)-Dihydro-6-methoxyterrecyclic acid A **166**	Weakly cytotoxic	Wijeratne et al. (2003)
A. terreus Thom	Xerophile	(+)-5(6)-Dihydro-6-hydroxyterrecyclic acid A **167**	Weakly cytotoxic	Wijeratne et al. (2003)
A. flavipes, from the rhizosphere of *Ericameria laricifolia* Nutt	Xerophile	Aspochalasins I–K (**168–170**)	Cytotoxic	Zhou et al. (2004)
A. terreus Thom from the rhizosphere of *Ambrosia ambrosioides* (Cav.) Payne	Xerophile	Terrequinone A **171**	Cytotoxic	He et al. (2004)
A. terreus Thom from the rhizosphere of *A. ambrosioides* (Cav.) Payne	Xerophile	Terrefuranone **172**	N/A	He et al. (2004)
A. cervinusi from the rhizosphere of *Anisacanthus thurberi* (Torr.) Gray	Xerophile	4*R*,5*S*-Dihydroxy-3-methoxy-5-methylcyclohex-2-enone **173**	N/A	He et al. (2004)
A. cervinusi from the rhizosphere of *A. thurberi* (Torr.) Gray	Xerophile	6-Methoxy-5(6)-dihydropenicillic acid **174**	N/A	He et al. (2004)
Chaetomium globosum	Xerophile	Globosumones A–C (**175–177**)	Cytotoxic	Bashyal et al. (2005)
A. restrictus strain A-17; *A. pennicilliodes* SPEG	Xerophile	Arestrictins A and B (**178–179**)	N/A	Huang et al. (2009)
Geopyxis aff. *Majalis* (**180–183**); *Geopyxis* sp. AZ0066 (**184–185**)	Xerophile	Geopyxins A–F (**180–185**)	Cytotoxic	Scott et al. (2008), Oh et al. (2009)

(continued)

Table 7 (continued)

Extremophile	Classification	Compound	Bioactivity	References
Sphingomonas astaxanthinifaciens strain TDMA-17	Radiotolerant/ toxitolerant	Astaxanthin dirhamnoside **186**	Antioxidant	Asker et al. (2009)
Streptomyces sp. RM-4-15	Metallotolerant	Frenolicins C–G (**187–191**)	N/A	Wang et al. (2013)
Streptomyces sp. RM-7-15	Metallotolerant	Herbimycins D–F (**192–194**)	Nontoxic Hsp90 inhibitor	Shaaban et al. (2013)
S. sannanesis	Unclassifiable	Sannastatin **195**	Toxic	Yang et al. (2011)
S. sannanesis	Unclassifiable	Vicenistatin **196**	Cytotoxic	Yang et al. (2011)
S. violascens YIM 100525 (**197–203**); *Streptomyces* strain 112CH148 (**204–205**)	Unclassifiable	Violapyrones A–I (**197–205**)	Antimicrobial (**197–203**); cytotoxic	Zhang et al. (2013), Shin et al. (2014)

N/A information not available

a various diseases at the individual and population levels. Ubiquitous microbiota, including those of the genera *Clostridia*, *Proprionibacteria*, and *Burkholderia*, have the biosynthetic capacity to produce secondary metabolites that interact with organisms within their environment to trigger immune responses, such as the production of chemokines in the host. With further development of analytical, biochemical, and bioinformatic workflows, including new mass spectrometry and nuclear magnetic resonance techniques, the metabalome of animal microbiomes will be cataloged and we will have more tools identify new bioactive metabolites from unique communities of microorganisms. For a recent article on finding new antibiotics from the vaginal commensal, see the paper by Donia et al. (2014).

Table 7 summarizes the bioactive agents isolated from extremophiles that are rock associated, toxitolerant, metallotolerant, and those deemed as unclassified. Once there are more studies on the isolation microorganisms from these unique environmental niches and characterization of their metabolomes, more reports of new bioactive chemical entities from these extremophiles will be published.

9 Summary and Concluding Remarks

As the search for new phamacophores continues, researchers around the world are now thoroughly investigating the metabolomes of extremophiles inhabiting unique geological niches. However, the legal issues, such as international treaties, national legislation, and self-regulation, related to bioprospecting have made collecting biologically active agents from source (host) countries more challenging than before (Kinghorn 2001; Cragg et al. 2012). The United Nations Convention on Biological Diversity (CBD) held in Rio de Janiero in 1992 led to >98 % of all countries, with the notable exception of the United States, ratifying an international treaty to conserve biodiversity, sustain the use of biodiversity, and share the benefits of using genetic resources. Although this treaty created a framework for fair access and benefit sharing (ABS) regarding genetic resources, it has only been partly successful in clarifying the legal issues related to natural product discovery. Subsequent revisions have been made to create a fair and equitable framework for ABS between host countries and researchers. For more details on the impact of the CBD on natural products research, see the 2012 review by Cragg and coworkers (Cragg et al. 2012).

In addition to legal issues, the lack of technological advances in underdeveloped host countries has also imposed difficulties in bioprospecting. For example, the first microbial natural product was reported only 14 years ago from Brazil, a country with a variety of unique, biologically diversity-rich ecological niches (Ióca et al. 2014). However, the 2014 review by Ióca, Allard, and Berlinck discusses the slow increase in the number of bioactive microbial metabolites that has happened as a result of the development of genetic and molecular biology tools as well as strong (inter)national collaborations. This trend is expected to continue in Brazil as well as many other countries with the development of new technology, multidisciplinary research collaborations, and increasing number of FDA-approved drugs and bioactive agents in human clinical trials derived from extreme environments.

With the development of next-generation sequencing technologies and the increasing availability of genetic information, we can now make informed decisions about how to characterize and fine tune the expression of genes, molecular interactions, cross-species gene expression, signal transduction, as well as activate and silence various genes involved in the production of bioactive secondary metabolites. The development of new bioassays and high-throughput screening methods will be the next frontier that needs to be explored to determine whether isolated metabolites have two or more different bioactivities. With big pharmaceutical companies currently not investing in natural product discovery efforts, new, inexpensive biological screens need to be developed to identify novel chemical agents. Furthermore, other initiatives have been started to repurpose known bioactive agents. At the National Institutes of Health, the National Center for Advancing Translational Sciences (NCATS; http://www.ncats.nih.gov/) has begun retesting FDA-approved and investigational agents in different biological screens to find other new drugs for the treatment of Alzheimer's disease, cancer, malaria, fungal diseases, and mycobacterial infections. Thus, increased access to more high-throughput screens will be essential to fully explore the bioactivity of small molecules and identify useful pharmacophores that will aid in the development of new, selective drugs.

References

Ahmed FR, Buckingham MJ, Hawkes GE, Toube TP (1984) Pyrrolylpolyenes. V: revision of the structures of the principal pigments of *Wallemia sebi*. A nuclear Overhauser enhancement study. J Chem Res, Synop 6:178–179

Aislabie JM, Balks MR, Foght JM, Waterhouse EJ (2004) Hydrocarbon spills on Antarctic soils: effects and management. Environ Sci Technol 38:1265–1274

Al-Zereini W, Schuhmann I, Laatsch H, Helmke E, Anke H (2007) New aromatic nitro compounds from *Salegentibacter* sp. T436, an arctic sea ice bacterium: taxonomy, fermentation, isolation and biological activities. J Antibiot 60:301–308

Alberola A, Meseguer I, Torreblanca M, Moya A, Sancho S, Polo B, Soria B, Such L (1998) Halocin H7 decreases infarct size and ectopic beats after myocardial reperfusion in dogs. J Physiol 509:148P

Alfredson DA, Akhurst RJ, Korolik V (2003) Antimicrobial resistance and genomic screening of clinical isolates of thermophilic *Campylobacter* spp. from south-east Queensland. Aust J Appl Microbiol 94:495–500

Alley SC, Okeley NM, Senter PD (2010) Antibody-drug conjugates: targeted drug delivery for cancer. Curr Opin Chem Biol 14:529–537

Aman S, Anderson DJ, Connolly TJ, Crittall AJ, Ji G (2000) From adenosine to 3'-deoxy-adenosine: development and scale up. Org Proc Res Dev 4:601–605

Andrioli WJ, Santos MS, Silva VB, Oliveira RB, Chagas-Paula DA, Jorge JA, Furtado NAJC, Pupo MT, Silva CHTP, Naal RMZG (2012) δ-Lactam derivative from thermophilic soil fungus exhibits in vitro anti-allergic activity. Nat Prod Res 26:2168–2175

Aragozzini F, Manachini PL, Craveri R, Rindone B, Scolastico C (1972) Structure of thermozy-mocidin. Experientia 28:881–882

Asker D, Amano S-I, Morita K, Tamura K, Sakuda S, Kikuchi N, Furihata K, Matsufuji H, Beppu T, Ueda K (2009) Astaxanthin dirhamnoside, a new astaxanthin derivative produced by a radio-tolerant bacterium, *Sphingomonas astaxanthinifaciens*. J Antibiot 62:397–399

Asker D, Awad TS, Beppu T, Ueda K (2012) Isolation, characterization, and diversity of novel radiotolerant carotenoid-producing bacteria. In: Barredo J-L (ed) Microbial carotenoids from bacteria and microalgae, vol 892. Methods in molecular biology. Humana Press, New York, pp 21–60

Badji B, Riba A, Mathieu F, Lebrihi A, Sabaou N (2005) Activité antifongique d'une souche d'*Actinomadura* d'origine saharienne sur divers champignons pathogènes et toxinogènes. J Med Mycol 15:211–219

Bagli J, Kluepfel D, St.-Jacques M (1973) Elucidation of structure and stereochemistry of myrio-cin. Novel antifungal antibiotic. J Org Chem 38:1253–1260

© The Author(s) 2015

L.-A. Giddings and D.J. Newman, *Bioactive Compounds from Terrestrial Extremophiles*, Extremophilic Bacteria, DOI 10.1007/978-3-319-13260-0

Bahn Y, Park J, Bai D, Takase S, Yu J (1998) YUA001, a novel aldose reductase inhibitor isolated from alkalophilic *Corynebacterium* sp. YUA25. I. Taxonomy, fermentation, isolation and characterization. J Antibiot 51:902–907

Balibar CJ, Howard-Jones AR, Walsh CT (2007) Terrequinone A biosynthesis through L-tryptophan oxidation, dimerization and bisprenylation. Nat Chem Biol 3:584–592

Ball P, Stillinger FH (1999) H_2O: a biography of water. Nature 401:850

Banciu H, Sorokin DY, Rijpstra WIC, Sinninghe Damsté JS, Galinski EA, Takaichi S, Muyzer G, Kuenen JG (2005) Fatty acid, compatible solute and pigment composition of obligately chemolithoautotrophic alkaliphilic sulfur-oxidizing bacteria from soda lakes. FEMS Microbiol Lett 243:181–187

Bandara WMMS, Seneviratne G, Kulasooriya SA (2006) Interactions among endophytic bacteria and fungi: effects and potentials. J Biosci 31:645–650

Banerjee M, Everroad RC, Castenholz RW (2009) An unusual cyanobacterium from saline thermal waters with relatives from unexpected habitats. Extremophiles 13:707–716

Barns SM, Fundyga RE, Jeffries MW, Pace NR (1994) Remarkable archaeal diversity detected in a Yellowstone National Park hot spring environment. Proc Nat Acad Sci USA 91:1609–1613

Bashyal BP, Wijeratne EMK, Faeth SH, Gunatilaka AAL (2005) Globosumones A-C, cytotoxic orsellinic acid esters from the Sonoran Desert endophytic fungus *Chaetomium globosum*. J Nat Prod 68:724–728

Bender CF, Yoshimoto FK, Paradise CL, Brabander JKD (2009) A concise synthesis of berkelic acid inspired by combining the natural products spicifernin and pulvilloric acid. J Am Chem Soc 131:11350–11352

Bentley H, Cunningham K, Spring F (1951) 509. Cordycepin, a metabolic product from cultures of *Cordyceps militaris* (Linn.) link. Part II. The structure of cordycepin. J Chem Soc 1951:2301–2305

Berdeja JG (2013) Lorvotuzumab mertansine: antibody-drug-conjugate for CD56+ multiple myeloma. Front biosci (Landmark Ed) 19:163–170

Bertrand S, Bohni N, Schnee S, Schumpp O, Gindro K, Wolfender J-L (2014) Metabolite induction via microorganism co-culture: a potential way to enhance chemical diversity for drug discovery. Biotechnol Adv 32:1180–1204

Bidasee KR, Nallani K, Besch HR, Dincer UD (2003) Streptozotocin-induced diabetes increases disulfide bond formation on cardiac ryanodine receptor (RyR2). J Pharmacol Exp Ther 305:989–998

Bissett D, Graham M, Setanoians A, Chadwick G, Wilson P, Koier I, Henrar R, Schwartsmann G, Cassidy J, Kaye S (1992) Phase I and pharmacokinetic study of rhizoxin. Cancer Res 52:2894–2898

Bolhuis H, Stal LJ (2011) Analysis of bacterial and archaeal diversity in coastal microbial mats using massive parallel 16S rRNA gene tag sequencing. ISME J 5:1701–1712

Bomar L, Maltz M, Colston S, Graf J (2011) Directed culturing of microorganisms using metatranscriptomics. mBio 2(2):e00012–e00011

Botić T, Kunčič MK, Sepčić K, Knez Ž, Gunde-Cimerman N (2012) Salt induces biosynthesis of hemolytically active compounds in the xerotolerant food-borne fungus *Wallemia sebi*. FEMS Microbiol Lett 326:40–46

Boudjella H, Bouti K, Zitouni A, Mathieu F, Lebrihi A, Sabaou N (2006) Taxonomy and chemical characterization of antibiotics of *Streptosporangium* Sg 10 isolated from a Saharan soil. Microbiol Res 161:288–298

Bouhired S, Weber M, Kempf-Sontag A, Keller NP, Hoffmeister D (2007) Accurate prediction of the *Aspergillus nidulans* terrequinone gene cluster boundaries using the transcriptional regulator LaeA. Fungal Genet Biol 44:1134–1145

Brimble M, Nairn M, Duncalf L (1999) Pyranonaphthoquinone antibiotics—isolation, structure and biological activity. Nat Prod Rep 16:267–281

Bruntner C, Binder T, Pathom-aree W, Goodfellow M, Bull AT, Potterat O, Puder C, Horer S, Schmid A, Bolek W, Wagner K, Mihm G, Fiedler H-P (2005) Frigocyclinone, a novel angucyclinone antibiotic produced by a *Streptomyces griseus* strain from Antarctica. J Antibiot 58:346–349

Buchgraber P, Snaddon TN, Wirtz C, Mynott R, Goddard R, Fürstner A (2008) A synthesis-driven structure revision of berkelic acid methyl ester. Angew Chem Int Ed 47:8450–8454

Bulgarelli D, Rott M, Schlaeppi K, Ver Loren van Themaat E, Ahmadinejad N, Assenza F, Rauf P, Huettel B, Reinhardt R, Schmelzer E, Peplies J, Gloeckner FO, Amann R, Eickhorst T, Schulze-Lefert P (2012) Revealing structure and assembly cues for *Arabidopsis* root-inhabiting bacterial microbiota. Nature 488:91–95

Bulkley D, Johnson F, Steitz TA (2012) The antibiotic thermorubin inhibits protein synthesis by binding to inter-subunit bridge B2a of the ribosome. J Mol Biol 416:571–578

Busti E, Monciardini P, Cavaletti L, Bamonte R, Lazzarini A, Sosio M, Donadio S (2006) Antibiotic-producing ability by representatives of a newly discovered lineage of actinomycetes. Microbiology 152:675–683

Buzzini P, Branda E, Goretti M, Turchetti B (2012) Psychrophilic yeasts from worldwide glacial habitats: diversity, adaptation strategies and biotechnological potential. FEMS Microbiol Ecol 82:217–241

Cassady JM, Chan KK, Floss HG, Leistner E (2004) Recent developments in the maytansinoid antitumor agents. Chem Pharm Bull 52:1–26

Cavalleri B, Turconi M, Pallanza R (1985) Synthesis and antibacterial activity of some derivatives of the antibiotic thermorubin. J Antibiot 38:1752–1760

Chandra S (2012) Endophytic fungi: novel sources of anticancer lead molecules. Appl Microbiol Biotechnol 95:47–59

Chen JK, Lane WS, Schreiber SL (1999) The identification of myriocin-binding proteins. Chem Biol 6:221–235

Chen L, Wang G, Bu T, Zhang Y, Wang Y, Liu M, Lin X (2010) Phylogenetic analysis and screening of antimicrobial and cytotoxic activities of moderately halophilic bacteria isolated from the Weihai Solar Saltern (China). World J Microbiol Biotechnol 26:879–888

Chu Y-S, Niu X-M, Wang Y-L, Guo J-P, Pan W-Z, Huang X-W, Zhang K-Q (2010) Isolation of putative biosynthetic intermediates of prenylated indole alkaloids from a thermophilic fungus *Talaromyces thermophilus*. Org Lett 12:4356–4359

Churro C, Alverca E, Sam-Bento F, Paulino Sr, Figueira V, Bento A, Prabhakar S, Lobo A, Calado A, Pereira P (2009) Effects of bacillamide and newly synthesized derivatives on the growth of cyanobacteria and microalgae cultures. J Appl Phycol 21:429–442

Clark BR, Capon RJ, Lacey E, Tennant S, Gill JH (2006) Polyenylpyrroles and polyenylfurans from an Australian isolate of the soil ascomycete *Gymnoascus reessii*. Org Lett 8:701–704

Corey EJ, Weigel LO, Chamberlin AR, Cho H, Hua DH (1980) Total synthesis of maytansine. J Am Chem Soc 102:6613–6615

Cowan DA, Makhalanyane TP, Dennis PG, Hopkins DW (2014) Microbial ecology and biogeochemistry of continental Antarctic soils. Front Microbiol 5:154

Cox CD, Rinehart KL, Moore ML, Cook JC (1981) Pyochelin: novel structure of an iron-chelating growth promoter for *Pseudomonas aeruginosa*. Proc Nat Acad Sci USA 78:4256–4260

Cragg GM, Katz F, Newman DJ, Rosenthal J (2012) The impact of the United Nations Convention on Biological Diversity on natural products research. Nat Prod Rep 29:1407–1423

Craveri R, Coronelli C, Pagani H, Sensi P (1964) Thermorubi, a new antibiotic from a thermo-actinomycete. Clin Med 71:511–521

Craveri R, Manachini PL, Aragozzini F (1972) Thermozymocidin new antifungal antibiotic from a thermophilic eumycete. Experientia 28:867–868

Cunningham K, Manson W, Spring F (1950) Cordycepin, a metabolic product isolated from cultures of *Cordyceps militaris* (Linn.) Link. Nature 166:949

D'Amico S, Collins T, Marx J-C, Feller G, Gerday C (2006) Psychrophilic microorganisms: challenges for life. EMBO Rep 7:385–389

Dalsgaard PW, Blunt JW, Munro MHG, Frisvad JC, Christophersen C (2005) Communesins G and H, new alkaloids from the psychrotolerant fungus *Penicillium rivulum*. J Nat Prod 68:258–261

Dalsgaard PW, Blunt JW, Munro MHG, Larsen TO, Christophersen C (2004) Psychrophilin B and C: cyclic nitropeptides from the psychrotolerant fungus *Penicillium rivulum*. J Nat Prod 67:1950–1952

Dalsgaard PW, Larsen TO, Christophersen C (2005) Bioactive cyclic peptides from the psychro-tolerant fungus *Penicillium algidum*. J Antibiot 58:141–144

Dalsgaard PW, Larsen TO, Frydenvang K, Christophersen C (2004) Psychrophilin A and cyclo-aspeptide D, novel cyclic peptides from the psychrotolerant fungus *Penicillium ribeum*. J Nat Prod 67:878–881

Darling CA, Siple PA (1941) Bacteria of Antarctica. J Bacteriol 42:83–98

De Maayer P, Anderson D, Cary C, Cowan DA (2014) Some like it cold: understanding the sur-vival strategies of psychrophiles. EMBO Rep 15:508–517

Dietera A, Hamm A, Fiedler H-P, Goodfellow M, Müller W, Brun R, Beil W, Bringmann G (2003) Pyrocoll, an antibiotic, antiparasitic and antitumor compound produced by a novel alkaliphilic *Streptomyces* strain. J Antibiot 56:639–646

Dimise EJ, Widboom PF, Bruner SD (2008) Structure elucidation and biosynthesis of fuscache-lins, peptide siderophores from the moderate thermophile *Thermobifida fusca*. Proc Nat Acad Sci USA 105:15311–15316

Ding Z-G, Li M-G, Zhao J-Y, Ren J, Huang R, Xie M-J, Cui X-L, Zhu H-J, Wen M-L (2010) Naphthospironone A: an unprecedented and highly functionalized polycyclic metabolite from an alkaline mine waste extremophile. Chem Eur J 16:3902–3905

Ding Z-G, Zhao J-Y, Li M-G, Huang R, Li Q-M, Cui X-L, Zhu H-J, Wen M-L (2012) Griseusins F and G, spiro-naphthoquinones from a tin mine tailings-derived alkalophilic *Nocardiopsis* species. J Nat Prod 75:1994–1998

Donia MS, Cimermancic P, Schulze CJ, Wieland Brown LC, Martin J, Mitreva M, Clardy J, Linington RG, Fischbach MA (2014) A systematic analysis of biosynthetic gene clusters in the human microbiome reveals a common family of antibiotics. Cell 158:1402–1414

Dorn RI, Oberlander TM (1981) Microbial origin of desert varnish. Science 213:1245–1247

Du J (2014) Use of cordycepin in manufacture of medicaments for anti-depression. Patent Number WO 2014029285

Ebada SS, Fischer T, Hamacher A, Du F-Y, Roth YO, Kassack MU, Wang B-G, Roth EH (2014) Psychrophilin E, a new cyclotripeptide, from co-fermentation of two marine alga-derived fungi of the genus *Aspergillus*. Nat Prod Res 28:776–781

Eloff JN, Masoko P, Picard J (2007) Resistance of animal fungal pathogens to solvents used in bioassays. S Afr J Bot 73:667–669

Esikova TZ, Temirov YV, Sokolov SL, Alakhov YB (2002) Secondary antimicrobial metabo-lites produced by *Thermophilic Bacillus* spp. strains VK2 and VK21. App Environ Microbiol 38:226–231

Etoh T, Kim YP, Tanaka H, Hayashi M (2013) Anti-inflammatory effect of berkeleyacetal C through the inhibition of interleukin-1 receptor-associated kinase-4 activity. Eur J Pharmacol 698:435–443

Fernández AB, Ghai R, Martin-Cuadrado A-B, Sánchez-Porro C, Rodriguez-Valera F, Ventosa A (2014) Prokaryotic taxonomic and metabolic diversity of an intermediate salinity hypersaline habitat assessed by metagenomics. FEMS Microbiol Ecol 88:623–635

Fernández AB, Vera-Gargallo B, Sánchez-Porro C, Ghai R, Papke RT, Rodriguez-Valera F, Ventosa A (2014) Comparison of prokaryotic community structure from Mediterranean and Atlantic saltern concentrator ponds by a metagenomic approach. Front Microbiol 5:196

Fish SA, Codd GA (1994) Bioactive compound production by thermophilic and thermotolerant cyanobacteria (blue-green algae). World J Microbiol Biotechnol 10:338–341

Fleck WF, Strauss DG, Prauser H (1980) Naphthoquinone antibiotics from Streptomyces lat-eritius. I Fermentation, isolation and characterization of granatomycins A, C, and D. Z Allg Mikrobiol 20:543–551

Foster J (1887) Über eihnigh Eigenschafter leuchtender Bakterien. Centr Bakteriol Rev Oarasitenk 2:337–340

Foti M, Sorokin D, Zacharova E, Pimenov N, Kuenen JG, Muyzer G (2008) Bacterial diversity and activity along a salinity gradient in soda lakes of the Kulunda Steppe (Altai, Russia). Extremophiles 12:133–145

Frisvad JC, Larsen TO, Dalsgaard PW, Seifert KA, Louis-Seize G, Lyhne EK, Jarvis BB, Fettinger JC, Overy DP (2006) Four psychrotolerant species with high chemical diversity consistently producing cycloaspeptide A, *Penicillium jamesonlandense* sp. nov., *Penicillium ribium* sp. nov., *Penicillium soppii* and *Penicillium lanosum*. Int J Syst Evol Microbiol 56:1427–1437

Fu X, Schmitz FJ, Tanner RS (1995) Chemical constituents of halophilic facultatively anaerobic bacteria, 1. J Nat Prod 58:1950–1954

Fujita T, Inoue K, Yamamoto S, Ikumoto T, Sasaki S, Toyama R, Chiba K, Hoshino Y, Okumoto T (1994) Fungal metabolites. Part 11. A potent immunosuppressive activity found in *Isaria sinclairii* metabolite. J Antibiot 47:208–215

Gabani P, Singh O (2013) Radiation-resistant extremophiles and their potential in biotechnology and therapeutics. Appl Microbiol Biotechnol 97:993–1004

Gai S, Zhang Q, Hu X (2014) Total synthesis of asterredione. J Org Chem 79:2111–2114

Gerday C, Glansdorff N (2007) Forward. In: Gerday C, Glansdorff N (eds) Physiology and biochemistry of extremophiles. ASM Press, Washinton, pp xi–xiii

Gesheva V, Vasileva-Tonkova E (2012) Production of enzymes and antimicrobial compounds by halophilic Antarctic *Nocardioides* sp. grown on different carbon sources. World J Microbiol Biotechnol 28:2069–2076

Giddings L-A, Newman DJ (2013) Microbial natural products: molecular blueprints for antitumor drugs. J Ind Microbiol Biotechnol 40:1181–1210

Goers L, Freemont P, Polizzi KM (2014) Co-culture systems and technologies: taking synthetic biology to the next level. J R Soc Interface 11:20140065

Gokul JK, Valverde A, Tuffin M, Cary SC, Cowan DA (2013) Micro-eukaryotic diversity in hypolithons from Miers Valley, Antarctica. Biology 2:331–340

Guerry P, Perez-Casal J, Yao R, McVeigh A, Trust TJ (1997) A genetic locus involved in iron utilization unique to some *Campylobacter* strains. J Bacteriol 179:3997–4002

Gunatilaka AAL (2006) Natural products from plant-associated microorganisms: distribution, structural diversity, bioactivity, and implications of their occurrence. J Nat Prod 69:509–526

Guo J-P, Tan J-L, Wang Y-L, Wu H-Y, Zhang C-P, Niu X-M, Pan W-Z, Huang X-W, Zhang K-Q (2011) Isolation of talathermophilins from the thermophilic fungus *Talaromyces thermophilus* YM3-4. J Nat Prod 74:2278–2281

Guo J-P, Zhu C-Y, Zhang C-P, Chu Y-S, Wang Y-L, Zhang J-X, Wu D-K, Zhang K-Q, Niu X-M (2012) Thermolides, potent nematocidal PKS-NRPS hybrid metabolites from thermophilic fungus *Talaromyces thermophilus*. J Am Chem Soc 134:20306–20309

Hafenbradl D, Keller M, Stetter KO, Hammann P, Hoyer F, Kogler H (1996) Sibyllimycine, 5,6,7,8-tetrahydro-3-methyl-8-oxo-4-azaindolizidine, a novel metabolite from *Thermoactinomyces* sp. Angew Chem Int Ed 35:545–547

Hamel E (1992) Natural products which interact with tubulin in the vinca domain: maytansine, rhizoxin, phomopsin a, dolastatins 10 and 15 and halichondrin B. Pharmacol Ther 55:31–51

Hansske F, Robins MJ (1985) Regiospecific and stereoselective conversion of ribonucleosides to 3'-deoxynucleosides. A high yield three-stage synthesis of cordycepin from adenosine. Tetrahedron Lett 26:4295–4298

Hayashi K-I, Dombou M, Sekiya M, Nakajima H, Fujita T, Nakayama M (1995) Thermorubin and 2-hydroxyphenyl acetic acid, aldose reductase inhibitors. J Antibiot 48:1345–1346

He J, Roemer E, Lange C, Huang X, Maier A, Kelter G, Jiang Y, Xu L-h, Menzel K-D, Grabley S, Fiebig H-H, Jiang C-L, Sattler I (2007) Structure, derivatization, and antitumor activity of new griseusins from *Nocardiopsis* sp. J Med Chem 50:5168–5175

He J, Wijeratne EMK, Bashyal BP, Zhan J, Seliga CJ, Liu MX, Pierson EE, Pierson LS, VanEtten HD, Gunatilaka AAL (2004) Cytotoxic and other metabolites of *Aspergillus* inhabiting the rhizosphere of Sonoran Desert plants. J Nat Prod 67:1985–1991

Hernandes MZ, Cavalcanti SMT, Moreira DRM, de Azevedo J, Filgueira W, Leite ACL (2010) Halogen atoms in the modern medicinal chemistry: hints for the drug design. Curr Drug Targets 11:303–314

Higashide E, Asai M, Ootsu K, Tanida S, Kozai Y, Hasegawa T, Kishi T, Sugino Y, Yoneda M (1977) Ansamitocin, a group of novel maytansinoid antibiotics with antitumour properties from *Nocardia*. Nature 270:721–722

Höfle G, Pohlan S, Uhlig G, Kabbe K, Schumacher D (1994) Keronopsins A and B, chemical defence substances of the marine ciliate *Pseudokeronopsis rubra* (protozoa): identification by ex vivo HPLC. Angew Chem Int Ed 33:1495–1497

Höltzel A, Dieter A, Schmid DG, Brown R, Goodfellow M, Beil W, Jung G, Fiedler H-P (2003) Lactonamycin Z, an antibiotic and antitumor compound produced by *Streptomyces sanglieri* strain AK 623. J Antibiot 56:1058–1061

Hong J, White JD (2004) The chemistry and biology of rhizoxins, novel antitumor macrolides from *Rhizopus chinensis*. Tetrahedron 60:5653–5681

Hoover RB, Pikuta EV (2009) Psychrophilic and psychrotolerant microbial extremophiles in polar environments. In: Bej AK, Aislabie J, Atlas RM (eds) Polar microbiology: the ecology, biodiversity and bioremediation potential of microorganisms in extremely cold environments. CRC Press, Boca Raton

Horikoshi K, Bull AT (2011) Prologue: definition, categories, distribution, origin and evolution, pioneering studies, and emerging fields of extremophiles. In: Horikoshi K (ed) Extremophiles handbook, vol 1. Springer, Tokyo

Hrouzek P, Tomek P, Lukešová A, Urban J, Voloshko L, Pushparaj B, Ventura S, Lukavský J, Štys D, Kopecký J (2011) Cytotoxicity and secondary metabolites production in terrestrial *Nostoc* strains, originating from different climatic/geographic regions and habitats: is their cytotoxicity environmentally dependent? Environ Toxicol 26:345–358

Huang S-X, Wang X-J, Yan Y, Wang J-D, Zhang J, Liu C-X, Xiang W-S, Shen B (2012) Neaumycin: a new macrolide from *Streptomyces* sp. NEAU-x211. Org Lett 14:1254–1257

Huang S-X, Zhao L-X, Tang S-K, Jiang C-L, Duan Y, Shen B (2009) Erythronolides H and I, new erythromycin congeners from a new halophilic actinomycete *Actinopolyspora* sp. YIM90600. Org Lett 11:1353–1356

Igarashi M, Chen W, Tsuchida T, Umekita M, Sawa T, Naganawa H, Hamada M, Takeuchi T (1995) 4'-Deacetyl-(-)-griseusins A and B, new naphthoquinone antibiotics from an actinomycete. J Antibiot 48:1502–1505

Igarashi Y, Yanase S, Sugimoto K, Enomoto M, Miyanaga S, Trujillo ME, Saiki I, Kuwahara S (2011) Lupinacidin C, an inhibitor of tumor cell invasion from *Micromonospora lupini*. J Nat Prod 74:862–865

Iizuka T, Fudou R, Jojima Y, Ogawa S, Yamanaka S, Inukai Y, Ojika M (2006) Miuraenamides A and B, novel antimicrobial cyclic depsipeptides from a new slightly halophilic myxobacterium: taxonomy, production, and biological properties. J Antibiot 59:385–391

Inskeep WP, Rusch DB, Jay ZJ, Herrgard MJ, Kozubal MA, Richardson TH, Macur RE, Hamamura N, deM Jennings R, Fouke BW (2010) Metagenomes from high-temperature chemotrophic systems reveal geochemical controls on microbial community structure and function. PLoS ONE 5:e9773

Ióca LP, Allard P-M, Berlinck RGS (2014) Thinking big about small beings–the (yet) underdeveloped microbial natural products chemistry in Brazil. Nat Prod Rep 31:646–675

Itabashi T, Matsuishi N, Hosoe T, Toyazaki N, Udagawa S-i, Imai T, Adachi M, Kawai K-i (2006) Two new dioxopiperazine derivatives, arestrictins A and B, isolated from *Aspergillus restrictus* and *Aspergillus penicilloides*. Chem Pharm Bull 54:1639–1641

Ito Y, Shibata T, Arita M, Sawai H, Ohno M (1981) Chirally selective synthesis of sugar moiety of nucleosides by chemicoenzymatic approach: L-and D-riboses, showdomycin, and cordycepin. J Am Chem Soc 103:6739–6741

Ivanova V, Kolarova M, Aleksieva K, Gräfe U, Dahse HÄ, Laatsch H (2007) Microbiaeratin, a new natural indole alkaloid from a *Microbispora aerata* strain, isolated from Livingston Island, Antarctica. Prep Biochem Biotechnol 37:161–168

Iwasaki S, Kobayashi H, Furukawa J, Namikoshi M, Okuda S, Sato Z, Matsuda I, Noda T (1984) Studies on macrocyclic lactone antibiotics. VII. Structure of a phytotoxin "rhizoxin" produced by *Rhizopus chinensis*. J Antibiot 37:354–362

Jadulco R, Edrada RA, Ebel R, Berg A, Schaumann K, Wray V, Steube K, Proksch P (2003) New communesin derivatives from the fungus *Penicillium* sp. derived from the Mediterranean sponge *Axinella verrucosa*. J Nat Prod 67:78–81

Jeong S-Y, Ishida K, Ito Y, Okada S, Murakami M (2003) Bacillamide, a novel algicide from the marine bacterium, *Bacillus* sp. SY-1, against the harmful dinoflagellate. Cochlodinium polykrikoides. Tetrahedron Lett 44:8005–8007

Johnson D (2007) Physiology and ecology of acidophilic microorganisms. In: Gerdes C, Glandsorff N (eds) Physiology and Biochemistry of Extremophiles. ASM press, Washington, pp 257–270

Johnson F, Chandra B, Iden CR, Naiksatam P, Kahen R, Okaya Y, Lin S-Y (1980) Thermorubin 1. Structure studies. J Am Chem Soc 102:5580–5585

Karentz D, McEuen FS, Land MC, Dunlap WC (1991) Survey of mycosporine-like amino-acid compounds in Antarctic marine organisms—potential protection from ultraviolet exposure. Mar Biol 108:157–166

Karthikeyan P, Bhat SG, Chandrasekaran M (2013) Halocin SH10 production by an extreme haloarchaeon *Natrinema* sp. BTSH10 isolated from salt pans of South India. Saudi J of Biol Sci 20:205–212

Kasting J (1993) Earth's early atmosphere. Science 259:920–926

Keshri J, Mishra A, Jha B (2013) Microbial population index and community structure in saline–alkaline soil using gene targeted metagenomics. Microbiol Res 168:165–173

Kim BD, Park JH, Whang GS, Kim J, Lee HJ (2013) Health care food and food additive comprising cordycepin showing inhibitory effect of osteoclastogenesis. Korean patent KR 2013060851

Kinghorn AD (2001) Pharmacognosy in the 21st century. J Pharm Pharmacol 53:135–148

Kohama Y, Iida K, Itoh S, Tsujikawa K, Mimura T (1996) Increase of cytochrome b mRNA by bis(2-hydroxyethyl) trisulfide in J774A.1 cells. Biol Pharm Bull 19:876–878

Kohama Y, Iida K, Semba T, Mimura T, Inada A, Tanaka K, Nakanishi T (1992) Studies on thermophile products. IV. Structural elucidation of cytotoxic substance, BS-1, derived from *Bacillus stearothermophilus*. Chem Pharm Bull 40:2210–2211

Kohama Y, Iida K, Tanaka K, Semba T, Itoh M, Teramoto T, Tsujikawa K, Mimura T (1993) Studies on thermophile products. VI. Activation of mouse peritoneal macrophages by bis (2-hydroxyethyl) trisulfide. Biol Pharm Bull 16:973–977

Kohama Y, Teramoto T, Iida K, Murayama N, Semba T, Tatebatake N, Kayamori Y, Tsujikawa K, Mimura T (1991) Isolation of a new antitumor substance from *Bacillus stearothermophilus*. Chem Pharm Bull 39:2468–2470

Komuro I, Sunazuka T, Akagawa KS, Yokota Y, Iwamoto A, ōmura S (2008) Erythromycin derivatives inhibit HIV-1 replication in macrophages through modulation of MAPK activity to induce small isoforms of C/EBPß. Proc Nat Acad Sci USA 105:12509–12514

Kröber M, Bekel T, Diaz NN, Goesmann A, Jaenicke S, Krause L, Miller D, Runte KJ, Viehöver P, Pühler A (2009) Phylogenetic characterization of a biogas plant microbial community integrating clone library 16S-rDNA sequences and metagenome sequence data obtained by 454-pyrosequencing. J Biotechnol 142:38–49

Kumar R, Patel D, Bansal D, Mishra S, Mohammed A, Arora R, Sharma A, Sharma R, Tripathi R (2010) Extremophiles: sustainable resource of natural compounds-extremolytes. In: Singh OV, Harvey SP (eds) Sustainable biotechnology. Springer, Netherlands, pp 279–294

Kümler I, Ehlers Mortensen C, Nielsen DL (2011) Trastuzumab emtansine. Tumor-activated prodrug (TAP) immunoconjugate, oncolytic. Drugs Future 36:825–835

Kupchan SM, Komoda Y, Branfman AR, Sneden AT, Court WA, Thomas GJ, Hintz HPJ, Smith RM, Karim A (1977) Tumor inhibitors. 122. The maytansinoids. Isolation, structural elucidation, and chemical interrelation of novel ansa macrolides. J Org Chem 42:2349–2357

Kupchan SM, Komoda Y, Court WA, Thomas GJ, Smith RM, Karim A, Gilmore CJ, Haltiwanger RC, Bryan RF (1972) Tumor inhibitors. LXXIII. Maytansine, a novel antileukemic ansa macrolide from *Maytenus ovatus*. J Am Chem Soc 94:1354–1356

Kusari S, Hertweck C, Spiteller M (2012) Chemical ecology of endophytic fungi: origins of secondary metabolites. Chem Biol 19:792–798

Kusari S, Pandey SP, Spiteller M (2013) Untapped mutualistic paradigms linking host plant and endophytic fungal production of similar bioactive secondary metabolites. Phytochemistry 91:81–87

Lacap DC, Warren-Rhodes KA, McKay CP, Pointing SB (2011) Cyanobacteria and chloroflexi-dominated hypolithic colonization of quartz at the hyper-arid core of the Atacama Desert, Chile. Extremophiles 15:31–38

Lambert J (2010) Antibody-maytansinoid conjugates: a new strategy for the treatment of cancer. Drugs Future 35:471

Lawton EM, Cotter PD, Hill C, Ross RP (2007) Identification of a novel two-peptide lantibiotic, haloduracin, produced by the alkaliphile *Bacillus halodurans* C-125. FEMS Microbiol Lett 267:64–71

Lequerica JL, O'Connor JE, Such L, Alberola A, Meseguer I, Dolz M, Torreblanca M, Moya A, Colom F, Soria B (2006) A halocin acting on Na+/H+ exchanger of *Haloarchaea* as a new type of inhibitor in NHE of mammals. J Physiol Biochem 62:253–262

Li L, Li D, Luan Y, Gu Q, Zhu T (2012) Cytotoxic metabolites from the Antarctic psychrophilic fungus *Oidiodendron truncatum*. J Nat Prod 75:920–927

Li Y, Sun B, Liu S, Jiang L, Liu X, Zhang H, Che Y (2008) Bioactive asterric acid derivatives from the Antarctic ascomycete fungus *Geomyces* sp. J Nat Prod 71:1643–1646

Li Y, Xiang H, Liu J, Zhou M, Tan H (2003) Purification and biological characterization of halocin C8, a novel peptide antibiotic from halobacterium strain AS7092. Extremophiles 7:401–407

Li Y-Q, Li M-G, Li W, Zhao J-Y, Ding Z-G, Cui X-L, Wen M-L (2007) Griseusin D, a new pyra-nonaphthoquinone derivative from a alkaphilic *Nocardiopsis* sp. J Antibiot 60:757–761

Liu H, Xiao L, Wei J, Schmitz J, Liu M, Wang C, Cheng L, Wu N, Chen L, Zhang Y, Lin X (2013) Identification of *Streptomyces* sp. nov. WH26 producing cytotoxic compounds isolated from marine solar saltern in China. World J Microbiol Biotechnol 29:1271–1278

Liu J, Lei Y, Wang F, Yi Y, Liu Y, Wang G (2011) Immunostimulatory activities of specific bacterial secondary metabolite of *Anoxybacillus flavithermus* strain SX-4 on carp, *Cyprinus carpio*. J Appl Microbiol 110:1056–1064

Lu C, Shen Y (2007) A novel ansamycin, naphthomycin K from *Streptomyces* sp. J Antibiot 60:649–653

Lu Y, Liu Y, Xu Z, Li H, Liu H, Zhu W (2012) Halogen bonding for rational drug design and new drug discovery. Expert Opin Drug Discov 7:375–383

Lu Z-Y, Lin Z-J, Wang W-L, Du L, Zhu T-J, Fang Y-C, Gu Q-Q, Zhu W-M (2008) Citrinin dimers from the halotolerant fungus *Penicillium citrinum* B-57. J Nat Prod 71:543–546

Lundberg DS, Lebeis SL, Paredes SH, Yourstone S, Gehring J, Malfatti S, Tremblay J, Engelbrektson A, Kunin V, Rio TGd, Edgar RC, Eickhorst T, Ley RE, Hugenholtz P, Tringe SG, Dangl JL (2012) Defining the core *Arabidopsis thaliana* root microbiome. Nature 488:86–90

Ma Y, Galinski EA, Grant WD, Oren A, Ventosa A (2010) Halophiles 2010: life in saline environments. App Environ Microbiol 76:6971–6981

Macelroy RD (1974) Some comments on the evolution of extremophiles. Biosystems 6:74–75

Mancinelli RL (2005) Halophiles: a terrestrial analog for life in brines on Mars. In: Gunde-Cimerman N, Oren A, Plemenitaš A (eds) Adaptation to Life at High Salt Concentrations in *Archaea, Bacteria*, and *Eukarya*. Springer, Netherlands, pp 137–147

Marmann A, Aly AH, Lin W, Wang B, Proksch P (2014) Co-cultivation–a powerful emerging tool for enhancing the chemical diversity of microorganisms. Mar Drugs 12:1043–1065

Matsumoto N, Tsuchida T, Maruyama M, Kinoshita N, Homma Y, Iinuma H, Sawa T, Hamada M, Takeuchi T, Heida N (1999) Lactonamycin, a new antimicrobial antibiotic produced by *Streptomyces rishiriensis* MJ773-88K4. I. Taxonomy, fermentation, isolation, physico-chemical properties and biological activities. J Antibiot 52:269–275

May JJ, Wendrich TM, Marahiel MA (2001) The dhb operon of *Bacillus subtilis* encodes the biosynthetic template for the catecholic siderophore 2,3-dihydroxybenzoate-glycine-threonine trimeric ester bacillibactin. J Biol Chem 276:7209–7217

McClean AL (1918) Bacteria of the ice and snow in Antartica. Nature 102:35–39

McDonald FE, Gleason MM (1995) Asymmetric syntheses of stavudine (d4T) and cordycepin by cycloisomerization of alkynyl alcohols to endocyclic enol ethers. Angew Chem Int Ed 34:350–352

McLeod H, Murray L, Wanders J, Setanoians A, Graham M, Pavlidis N, Heinrich B, ten Bokkel Huinink W, Wagener D, Aamdal S (1996) Multicentre phase II pharmacological evaluation of rhizoxin. Eortc early clinical studies (ECSG)/pharmacology and molecular mechanisms (PAMM) groups. Br J Cancer 74:1944–1948

Meklat A, Sabaou N, Zitouni A, Mathieu F, Lebrihi A (2011) Isolation, taxonomy, and antagonistic properties of halophilic actinomycetes in Saharan soils of Algeria. App Environ Microbiol 77:6710–6714

Meseguer I, Rodriguez-Valera F (1985) Production and purification of halocin H4. FEMS Microbiol Lett 28:177–182

Meseguer I, Rodríguez-Valera F, Ventosa A (1986) Antagonistic interactions among halobacteria due to halocin production. FEMS Microbiol Lett 36:177–182

Meseguer I, Torreblanca M, Konishi T (1995) Specific inhibition of the halobacterial Na^+/H^+ antiporter by halocin H6. J Biol Chem 270:6450–6455

Meyers A, Shaw C-C (1974) Studies directed toward the total synthesis of maytansine. The preparation and properties of the carbinolamide moiety. Tetrahedron Lett 15:717–720

Mojib N, Philpott R, Huang JP, Niederweis M, Bej AK (2010) Antimycobacterial activity in vitro of pigments isolated from Antarctic bacteria. Antonie van Leeuwenhoek 98:531–540

Morita RY (1975) Psychrophilic bacteria. Bacteriol Rev 39:144

Nakada M, Kobayashi S, Iwasaki S, Ohno M (1993) The first total synthesis of the antitumor macrolide rhizoxin: synthesis of the key building blocks. Tetrahedron Lett 34:1035–1038

Narasingarao P, Podell S, Ugalde JA, Brochier-Armanet C, Emerson JB, Brocks JJ, Heidelberg KB, Banfield JF, Allen EE (2012) De novo metagenomic assembly reveals abundant novel major lineage of *Archaea* in hypersaline microbial communities. ISME J 6:81–93

Nichols D, Bowman J, Sanderson K, Nichols CM, Lewis T, McMeekin T, Nichols PD (1999) Developments with Antarctic microorganisms: culture collections, bioactivity screening, taxonomy, PUFA production and cold-adapted enzymes. Curr Opin Biotech 10:240–246

Numata A, Takahashi C, Ito Y, Takada T, Kawai K, Usami Y, Matsumura E, Imachi M, Ito T, Hasegawa T (1993) Communesins, cytotoxic metabolites of a fungus isolated from a marine alga. Tetrahedron Lett 34:2355–2358

Nützmann H-W, Reyes-Dominguez Y, Scherlach K, Schroeckh V, Horn F, Gacek A, Schümann J, Hertweck C, Strauss J, Brakhage AA (2011) Bacteria-induced natural product formation in the fungus *Aspergillus nidulans* requires Saga/Ada-mediated histone acetylation. Proc Nat Acad Sci USA 108:14282–14287

O'Connor EM, Shand RF (2002) Halocins and sulfolobicins: the emerging story of archaeal protein and peptide antibiotics. J Ind Microbiol Biotechnol 28:23–31

Oh D-C, Poulsen M, Currie CR, Clardy J (2009) Dentigerumycin: a bacterial mediator of an ant-fungus symbiosis. Nat Chem Biol 5:391–393

Oh D-C, Scott JJ, Currie CR, Clardy J (2009) Mycangimycin, a polyene peroxide from a mutualist *Streptomyces* sp. Org Lett 11:633–636

Okuyama E, Yamazaki M, Kobayashi K, Sakurai T (1983) Paraherquonin, a new meroterpenoid from *Penicillium paraherquei*. Tetrahedron Lett 24:3113–3114

Oman TJ, van der Donk WA (2009) Insights into the mode of action of the two-peptide lantibiotic haloduracin. ACS Chem Biol 4:865–874

Ōmura S, Suzuki Y, Kitao C, Takahashi Y, Konda Y (1975) Isolation of a new sulfur containing basic substance from a *Thermoactinomyces* species. J Antibiot 28:609–610

Onda M, Konda Y (1978) Synthesis of the alkaloid from *Thermoactinomyces* species. Chem Pharm Bull 26:2167–2169

Oren A (2010) Industrial and environmental applications of halophilic microorganisms. Environ Technol 31:825–834

Otto TC, Scott JR, Kauffman MA, Hodgins SM, diTargiani RC, Hughes JH, Sarricks EP, Hamilton TA, Cerasoli DM (2013) Identification and characterization of novel catalytic bioscavengers of organophosphorus nerve agents. Chem Biol Interact 203:186–190

Park HB, Kim Y-J, Lee JK, Lee KR, Kwon HC (2012) Spirobacillenes A and B, unusual spiro-cyclopentenones from *Lysinibacillus fusiformis* KMC003. Org Lett 14:5002–5005

Partida-Martinez LP, Hertweck C (2005) Pathogenic fungus harbours endosymbiotic bacteria for toxin production. Nature 437:884–888

Partida-Martinez LP, Hertweck C (2007) A gene cluster encoding rhizoxin biosynthesis in "*Burkholderia rhizoxina*", the bacterial endosymbiont of the fungus *Rhizopus microsporus*. ChemBioChem 8:41–45

Partida-Martinez LP, Monajembashi S, Greulich K-O, Hertweck C (2007) Endosymbiont-dependent host reproduction maintains bacterial-fungal mutualism. Curr Biol 17:773–777

Pašic L, Velikonja BH, Ulrih NP (2008) Optimization of the culture conditions for the production of a bacteriocin from halophilic archaeon Sech7a. Prep Biochem Biotechnol 38:229–245

Paterson RRM (2008) *Cordyceps*–a traditional Chinese medicine and another fungal therapeutic biofactory? Phytochemistry 69:1469–1495

Peppel WJ, Signaigo FK (1946). Preparation of hydroxythiols. US Patent 2402665, 25JUN1946

Pflüger K, Müller V (2004) Transport of compatible solutes in extremophiles. J Bioenerg Biomembr 36:17–24

Phoebe CH, Bombie J, Albert FG, Van Tran K, Cabrera J, Correira HJ, Guo Y, Lindermuth J, Rauert N, Galbraith W, Selitrennikoff CP (2001) Extremophilic organisms as an unexplored source of antifungal compounds. J Antibiot 54:56–65

Pikuta EV, Hoover RB, Tang J (2007) Microbial extremophiles at the limits of life. Crit Rev Microbiol 33:183–209

Platas G, Meseguer I, Amils R (2002) Purification and biological characterization of halocin H1 from *Haloferax mediterranei* M2a. Int Microbiol 5:15–19

Pointing SB, Belnap J (2012) Microbial colonization and controls in dryland systems. Nat Rev Micro 10:551–562

Pomati F, Rossetti C, Manarolla G, Burns BP, Neilan BA (2004) Interactions between intracellular Na^+ levels and saxitoxin production in *Cylindrospermopsis raciborskii* T3. Microbiology 150:455–461

Price LB, Shand RF (2000) Halocin S8: a 36-amino-acid microhalocin from the haloarchaeal strain S8a. J Bacteriol 182:4951–4958

Price PB (2000) A habitat for psychrophiles in deep Antarctic ice. Proc Nat Acad Sci USA 97:1247–1251

Pride DT, Schoenfeld T (2008) Genome signature analysis of thermal virus metagenomes reveals Archaea and thermophilic signatures. BMC Genomics 9:420

Pyrek JS, Achmatowicz O Jr, Zamojski A (1977) Naphto- and anthraquinones of *Streptomyces thermoviolaceus* WR-141. Structures and model syntheses. Tetrahedron 33:673–680

Rademacher A, Zakrzewski M, Schlüter A, Schönberg M, Szczepanowski R, Goesmann A, Pühler A, Klocke M (2012) Characterization of microbial biofilms in a thermophilic biogas system by high-throughput metagenome sequencing. FEMS Microbiol Ecol 79:785–799

Rdest U, Sturm M (1987) Bacteriocins from halobacteria. Protein purification: micro to macro. Alan R. Liss, Inc., New York

Riley MA (1998) Molecular mechanisms of bacteriocin evolution. Annu Rev Genet 32:255–278

Rodriguez-Valera F, Juez G, Kushner D (1982) Halocins: salt-dependent bacteriocins produced by extremely halophilic rods. Can J Microbiol 28:151–154

Rothschild LJ, Mancinelli RL (2001) Life in extreme environments. Nature 409:1092–1101

Rusman Y (2006) Isolation of new secondary metabolites from sponge-associated and plant-derived fungi Ph.D. Thesis. University of Dusseldorf, Dusseldorf

Russell NJ (1989) Adaptive modifications in membranes of halotolerant and halophilic microorganisms. J Bioenerg Biomembr 21:93–113

Sabaou N, Boudjella H, Bennadji A, Mostefaoui A, Zitouni A, Lamari L, Bennadji H, Lefèbvre G, Germain P (1998) Les sols des oasis du Sahara algérien, source d'actinomycètes, rares producteurs d'antibiotiques. Sécheresse 9:147–153

Sakakibara M, Masuda M (2009) Mutant of *Cordyceps militaris* and method of culturing the mutant. Japan Patent 2009034045

Sakurai A, Masuda M (2013) Manufacturing and purification method of cordycepin. Japan Patent 2013111060

Sarethy IP, Saxena Y, Kapoor A, Sharma M, Sharma SK, Gupta V, Gupta S (2011) Alkaliphilic bacteria: applications in industrial biotechnology. J Ind Microbiol Biotechnol 38:769–790

Sato M, Beppu T, Arima K (1980) Properties and structure of a novel peptide antibiotic No. 1907. Agric Biol Chem 44:3037–3042

Scherlach K, Busch B, Lackner G, Paszkowski U, Hertweck C (2012) Symbiotic cooperation in the biosynthesis of a phytotoxin. Angew Chem Int Ed 124:9753–9756

Scherlach K, Partida-Martinez LP, Dahse H-M, Hertweck C (2006) Antimitotic rhizoxin derivatives from a cultured bacterial endosymbiont of the rice pathogenic fungus *Rhizopus microsporus*. J Am Chem Soc 128:11529–11536

Schöner TA, Fuchs SW, Schönau C, Bode HB (2014) Initiation of the flexirubin biosynthesis in *Chitinophaga pinensis*. Microb Biotechnol 7:232–241

Schroeter B, Green T, Kappen L, Seppelt R (1994) Carbon dioxide exchange at subzero temperatures. Field measurements on *Umbilicaria aprina* in Antarctica. Cryptogam Bot 4:233–241

Scott JJ, Oh D-C, Yuceer MC, Klepzig KD, Clardy J, Currie CR (2008) Bacterial protection of beetle-fungus mutualism. Science 322:63

Sepcic K, Zalar P, Gunde-Cimerman N (2010) Low water activity induces the production of bio-active metabolites in halophilic and halotolerant fungi. Mar Drugs 9:43–58

Shaaban KA, Wang X, Elshahawi SI, Ponomareva LV, Sunkara M, Copley GC, Hower JC, Morris AJ, Kharel MK, Thorson JS (2013) Herbimycins D-F, ansamycin analogues from *Streptomyces* sp. RM-7-15. J Nat Prod 76:1619–1626

Shand RF, Leyva KJ (2007) Peptide and protein antibiotics from the domain Archaea: halocins and sulfolobicins. In: Riley MA, Chavan MA (eds) Bacteriocins. Springer, Berlin, pp 93–109

Shekh RM, Singh P, Singh SM, Roy U (2011) Antifungal activity of Arctic and Antarctic bacteria isolates. Polar Biol 34:139–143

Shih C-J, Chen P-Y, Liaw C-C, Lai Y-M, Yang Y-L (2014) Bringing microbial interactions to light using imaging mass spectrometry. Nat Prod Rep 31:739–755

Shin HJ, Lee H-S, Lee JS, Shin J, Lee MA, Lee H-S, Lee Y-J, Yun J, Kang JS (2014) Violapyrones H and I, new cytotoxic compounds isolated from *Streptomyces sp.* associated with the marine starfish *Acanthaster planci*. Mar Drugs 12:3283–3291

Short JM (2000) Screening for novel bioactivities. US Patent 6030779, 29FEB2000

Short JM, Keller M (2005) Detecting activity of biopolymer; screen library of clones, encapsulate biocompatible substrate and in gel drop, monitor drop for preferential activity, detect adjustment in substrate, adjustment in substrate indicates active biopolymer. US Patent 6872526 B2, 29MAR2005

Siddaramappa S, Challacombe JF, DeCastro RE, Pfeiffer F, Sastre DE, Giménez MI, Paggi RA, Detter JC, Davenport KW, Goodwin LA (2012) A comparative genomics perspective on the genetic content of the alkaliphilic haloarchaeon *Natrialba magadii* ATCC 43099T. BMC Genomics 13:165

Singh J, Tripathi R, Thakur IS (2014) Characterization of endolithic cyanobacterial strain, *Leptolyngbya* sp. ISTCY101, for prospective recycling of CO_2 and biodiesel production. Bioresour Technol 166:345–352

Sinha RP, Häder D-P (2008) UV-protectants in cyanobacteria. Plant Sci 174:278–289

Snipes CE, Chang C-J, Floss HG (1979) Biosynthesis of the antibiotic granaticin. J Am Chem Soc 101:701–706

Socha AM, Long RA, Rowley DC (2007) Bacillamides from a hypersaline microbial mat bacterium. J Nat Prod 70:1793–1795

Stetter KO (1996) Hyperthermophilic procaryotes. FEMS Microbiol Rev 18:149–158

Stierle A, Stierle D (2013) Bioprospecting in the Berkeley Pit: the use of signal transduction enzyme inhibition assays to isolate bioactive secondary metabolites from the extremophilic fungi of an acid mine waste lake. In: A-u Rahman (ed) Studies in natural products chemistry, vol 39. Elsevier, Oxford, pp 1–45

Stierle AA, Stierle DB (2005) Bioprospecting in the Berkeley Pit: bioactive metabolites from acid mine waste extremophiles. In: A-u Rahman (ed) Studies in natural products chemistry, vol 32. Elsevier, Oxford

Stierle AA, Stierle DB, Girtsman T (2012) Caspase-1 inhibitors from an extremophilic fungus that target specific leukemia cell lines. J Nat Prod 75:344–350

Stierle AA, Stierle DB, Kelly K (2006) Berkelic acid, a novel spiroketal with selective anticancer activity from an acid mine waste fungal extremophile. J Org Chem 71:5357–5360

Stierle AA, Stierle DB, Patacini B (2008) The berkeleyamides, amides from the acid lake fungus *Penicillum rubrum*. J Nat Prod 71:856–860

Stierle DB, Stierle AA, Girtsman T, McIntyre K, Nichols J (2012) Caspase-1 and -3 inhibiting drimane sesquiterpenoids from the extremophilic fungus *Penicillium solitum*. J Nat Prod 75:262–266

Stierle DB, Stierle AA, Hobbs JD, Stokken J, Clardy J (2004) Berkeleydione and berkeleytrione, new bioactive metabolites from an acid mine organism. Org Lett 6:1049–1052

Stierle DB, Stierle AA, Patacini B (2007) The berkeleyacetals, three meroterpenes from a deep water acid mine waste *Penicillium*. J Nat Prod 70:1820–1823

Stierle DB, Stierle AA, Patacini B, McIntyre K, Girtsman T, Bolstad E (2011) Berkeleyones and related meroterpenes from a deep water acid mine waste fungus that inhibit the production of interleukin 1-β from induced inflammasomes. J Nat Prod 74:2273–2277

Strobel G, Daisy B (2003) Bioprospecting for microbial endophytes and their natural products. Microbiol Mol Biol Rev 67:491–502

Strobel G, Daisy B, Castillo U, Harper J (2004) Natural products from endophytic microorganisms. J Nat Prod 67:257–268

Sun W-S, Lee H-S, Park J-m, Yu J-H, Kim S-h, Kim J-H (2001) YUA001, a novel aldose reductase inhibitor isolated from alkalophilic *Corynebacterium* sp. YUA25. II. Chemical modification and biological activity. J Antibiot 54:827–830

Takaichi S, Maoka T, Akimoto N, Sorokin DY, Banciu H, Kuenen JG (2004) Two novel yellow pigments natronochrome and chloronatronochrome from the natrono(alkali)philic sulfur-oxidizing bacterium *Thialkalivibrio versutus* strain ALJ 15. Tetrahedron Lett 45:8303–8305

Takami H, Nakasone K, Takaki Y, Maeno G, Sasaki R, Masui N, Fuji F, Hirama C, Nakamura Y, Ogasawara N (2000) Complete genome sequence of the alkaliphilic bacterium *Bacillus halodurans* and genomic sequence comparison with *Bacillus subtilis*. Nucleic Acids Res 28:4317–4331

Temirov YV, Esikova TZ, Balashova TA, Vinokurov LM, Alakhov YB (2003) A catecholic siderophore produced by the thermoresistant *Bacillus licheniformis* VK21 strain. Russ J Bioorg Chem 29:597–604

Thumar JT, Dhulia K, Singh SP (2010) Isolation and partial purification of an antimicrobial agent from halotolerant alkaliphilic *Streptomyces aburaviensis* strain Kut-8. World J Microbiol Biotechnol 26:2081–2087

Tian S-Z, Pu X, Luo G, Zhao L-X, Xu L-H, Li W-J, Luo Y (2013) Isolation and characterization of new *p*-terphenyls with antifungal, antibacterial, and antioxidant activities from halophilic actinomycete *Nocardiopsis gilva* YIM 90087. J Agric Food Chem 61:3006–3012

Tiwari R, Kalra A, Darokar MP, Chandra M, Aggarwal N, Singh AK, Khanuja SPS (2010) Endophytic bacteria from *Ocimum sanctum* and their yield enhancing capabilities. Curr Microbiol 60:167–171

Tolcher AW, Aylesworth C, Rizzo J, Izbicka E, Campbell E, Kuhn J, Weiss G, Von Hoff DD, Rowinsky EK (2000) A phase I study of rhizoxin (NSC 332598) by 72-hour continuous intravenous infusion in patients with advanced solid tumors. Ann Oncol 11:333–338

Torreblanca M, Meseguer I, Rodríguez-Valera F (1989) Halocin H6, a bacteriocin from *Haloferax gibbonsii*. J Gen Microbiol 135:2655–2661

Torreblanca M, Meseguer I, Ventosa A (1994) Production of halocin is a practically universal feature of archaeal halophilic rods. Lett Appl Microbiol 19:201–205

Tsuji N, Kobayashi M, Terui Y, Tori K (1976) The structures of griseusins A and B, new isochromanquinone antibiotics. Tetrahedron 32:2207–2210

Tsujibo H, Sato T, lnui M, Yamamoto H, InamPhysiology and Biochemistry of Extremophilesori Y (1988) Intracellular accumulation of phenazine antibiotics production by an alkalophilic actinomycete. Agric Biol Chem 52:301–306

Tsuruo T, Oh-hara T, Iida H, Tsukagoshi S, Sato Z, Matsuda I, Iwasaki S, Okuda S, Shimizu F, Sasagawa K (1986) Rhizoxin, a macrocyclic lactone antibiotic, as a new antitumor agent against human and murine tumor cells and their vincristine-resistant sublines. Cancer Res 46:381–385

Tuli H, Sandhu S, Sharma AK (2014) Pharmacological and therapeutic potential of *Cordyceps* with special reference to cordycepin. 3. Biotech 4:1–12

Vellieux F, Madern D, Zaccai G, Ebel C (2007) Molecular adaptation to high salt. In: Gerday C, Glansdorff N (eds) Physiology and Biochemistry of Extremophiles. ASM Press, Washington, pp 240–253

Volkmann M, Gorbushina AA, Kedar L, Oren A (2006) Structure of euhalothece-362, a novel red-shifted mycosporine-like amino acid, from a halophilic cyanobacterium (*Euhalothece* sp.). FEMS Microbiol Lett 258:50–54

Vreeland RH, Rosenzweig WD, Powers DW (2000) Isolation of a 250 million-year-old halotolerant bacterium from a primary salt crystal. Nature 407:897–900

Wächtershäuser G (2006) From volcanic origins of chemoautotrophic life to Bacteria, Archaea and Eukarya. Philos Trans R Soc B: Biol Sci 361:1787–1808

Waksman SA, Joffe JS (1922) Microorganisms concerned in the oxidation of sulfur in the soil: II. *Thiobacillus thiooxidans*, a new sulfur-oxidizing organism isolated from the soil 1. J Bacteriol 7:239

Wang G-X, Wang Y, Wu Z-F, Jiang H-F, Dong R-Q, Li F-Y, Liu X-L (2011) Immunomodulatory effects of secondary metabolites from thermophilic *Anoxybacillus kamchatkensis* XA-1 on carp, Cyprinus carpio. Fish Shellfish Immunol 30:1331–1338

Wang H, Wang Y, Wang W, Fu P, Liu P, Zhu W (2011) Anti-influenza virus polyketides from the acid-tolerant fungus *Penicillium purpurogenum* JS03-21. J Nat Prod 74:2014–2018

Wang H, Zheng J-K, Qu H-J, Liu P-P, Wang Y, Zhu W-M (2011) A new cytotoxic indole-3-ethenamide from the halotolerant fungus *Aspergillus sclerotiorum* PT06-1. J Antibiot 64:679–681

Wang W, Zhu T, Tao H, Lu Z, Fang Y, Gu Q, Zhu W (2007) Two new cytotoxic quinone type compounds from the halotolerant fungus *Aspergillus variecolor*. J Antibiot 60:603–607

Wang W-L, Lu Z-Y, Tao H-W, Zhu T-J, Fang Y-C, Gu Q-Q, Zhu W-M (2007) Isoechinulin-type alkaloids, variecolorins A-L, from halotolerant *Aspergillus variecolor*. J Nat Prod 70:1558–1564

Wang X, Shaaban KA, Elshahawi SI, Ponomareva LV, Sunkara M, Zhang Y, Copley GC, Hower JC, Morris AJ, Kharel MK, Thorson JS (2013) Frenolicins C-G, pyranonaphthoquinones from *Streptomyces* sp. RM-4-15. J Nat Prod 76:1441–1447

Wang Y, Zheng J, Liu P, Wang W, Zhu W (2011) Three new compounds from *Aspergillus terreus* PT06-2 grown in a high salt medium. Mar Drugs 9:1368–1378

Wijeratne EMK, Bashyal BP, Liu MX, Rocha DD, Gunaherath GMKB, U'Ren JM, Gunatilaka MK, Arnold AE, Whitesell L, Gunatilaka AAL (2012) Geopyxins A-E, ent-kaurane diterpenoids from endolichenic fungal strains *Geopyxis* aff. *majalis* and *Geopyxis* sp. AZ0066: structure–activity relationships of geopyxins and their analogues. J Nat Prod 75:361–369

Wijeratne EMK, Turbyville TJ, Zhang Z, Bigelow D, Pierson LS, VanEtten HD, Whitesell L, Canfield LM, Gunatilaka AAL (2003) Cytotoxic constituents of *Aspergillus terreus* from the rhizosphere of *Opuntia versicolor* of the Sonoran Desert. J Nat Prod 66:1567–1573

Wilkinson RA, Strobel G, Stierle A (1999) Sphaeric acid, a new succinic acid derivative from a *Sphaeropsis* sp. J Nat Prod 62:358–360

Wings S, Müller H, Berg G, Lamshöft M, Leistner E (2013) A study of the bacterial community in the root system of the maytansine containing plant *Putterlickia verrucosa*. Phytochemistry 91:158–164

Wong FY, Lacap D, Lau MY, Aitchison JC, Cowan D, Pointing S (2010) Hypolithic microbial community of quartz pavement in the high-altitude tundra of Central Tibet. Microb Ecol 60:730–739

Wu X, Zhou J, Snider BB (2009) Synthesis of (-)-berkelic acid. Angew Chem Int Ed 121:1309–1312

Wynn-Williams DD (1990) Ecological aspects of Antarctic microbiology. In: Marshall KC (ed) Advances in Microbial Ecology. Springer, New York, pp 71–146

Yamagishi Y, Matsuoka M, Odagawa A, Kato S, Shindo K, Mochizuki J (1993) Rumbrin, a new cytoprotective substance produced by *Auxarthron umbrinum*. I. Taxonomy, production, isolation and biological activities. J Antibiot 46:884–887

Yang S-W, Chan T-M, Terracciano J, Patel R, Patel M, Gullo V, Chu M (2006) A new hydrogenated azaphilone Sch 725680 from *Aspergillus* sp. J Antibiot 59:720–723

Yang S-X, Gao J-M, Zhang A-L, Laatsch H (2011) Sannastatin, a novel toxic macrolactam polyketide glycoside produced by actinomycete *Streptomyces sannanensis*. Bioorg Med Chem Lett 21:3905–3908

Yang Y-L, Liao W-Y, Liu W-Y, Liaw C-C, Shen C-N, Huang Z-Y, Wu S-H (2009) Discovery of new natural products by intact-cell mass spectrometry and LC-SPE-NMR: malbranpyrroles, novel polyketides from thermophilic fungus *Malbranchea sulfurea*. Chem Eur J 15:11573–11580

Zengler K, Toledo G, Rappé M, Elkins J, Mathur EJ, Short JM, Keller M (2002) Cultivating the uncultured. Proc Nat Acad Sci USA 99:15681–15686

Zhang J, Gallery M, Wyant T, Stringer B, Manfredi M, Danaee H, Veiby P (2013a) Abstract PR12: MLN0264, an investigational, first-in-class antibody-drug conjugate (ADC) targeting guanylyl cyclase C (GCC), demonstrates antitumor activity alone and in combination with gemcitabine in human pancreatic cancer xenograft models expressing GCC. Mol Cancer Ther 12:Abst PR12

Zhang J, Jiang Y, Cao Y, Liu J, Zheng D, Chen X, Han L, Jiang C, Huang X (2013) Violapyrones A-G, α-pyrone derivatives from Streptomyces violascens isolated from *Hylobates hoolock* feces. J Nat Prod 76:2126–2130

Zhang X, Alemany LB, Fiedler H-P, Goodfellow M, Parry RJ (2008) Biosynthetic investigations of lactonamycin and lactonamycin Z: cloning of the biosynthetic gene clusters and discovery of an unusual starter unit. Antimicrob Agents Chemother 52:574–585

Zhao L-X, Huang S-X, Tang S-K, Jiang C-L, Duan Y, Beutler JA, Henrich CJ, McMahon JB, Schmid T, Blees JS, Colburn NH, Rajski SR, Shen B (2011) Actinopolysporins A-C and tubercidin as a Pdcd4 stabilizer from the halophilic actinomycete *Actinopolyspora erythraea* YIM 90600. J Nat Prod 74:1990–1995

Zheng J, Wang Y, Wang J, Liu P, Li J, Zhu W (2013) Antimicrobial ergosteroids and pyrrole derivatives from halotolerant *Aspergillus flocculosus* PT05-1 cultured in a hypersaline medium. Extremophiles 17:963–971

Zheng L, Li G, Wang X, Pan W, Li L, Hua L, Liu F, Dang L, Mo M, Zhang K (2008) Nematicidal endophytic bacteria obtained from plants. Ann Microbiol 58:569–572

Zhou G-X, Wijeratne EMK, Bigelow D, Pierson LS, VanEtten HD, Gunatilaka AAL (2004) Aspochalasins I, J, and K: three new cytotoxic cytochalasans of *Aspergillus flavipes* from the rhizosphere of *Ericameria laricifolia* of the Sonoran Desert. J Nat Prod 67:328–332

Zitouni A, Boudjella H, Lamari L, Badji B, Mathieu F, Lebrihi A, Sabaou N (2005) *Nocardiopsis* and *Saccharothrix* genera in Saharan soils in Algeria: isolation, biological activities and partial characterization of antibiotics. Res Microbiol 156:984–993